T0142482

Springer Theses

Recognizing Outstanding Ph.D. Research

Aims and Scope

The series "Springer Theses" brings together a selection of the very best Ph.D. theses from around the world and across the physical sciences. Nominated and endorsed by two recognized specialists, each published volume has been selected for its scientific excellence and the high impact of its contents for the pertinent field of research. For greater accessibility to non-specialists, the published versions include an extended introduction, as well as a foreword by the student's supervisor explaining the special relevance of the work for the field. As a whole, the series will provide a valuable resource both for newcomers to the research fields described, and for other scientists seeking detailed background information on special questions. Finally, it provides an accredited documentation of the valuable contributions made by today's younger generation of scientists.

Theses are accepted into the series by invited nomination only and must fulfill all of the following criteria

- They must be written in good English.
- The topic should fall within the confines of Chemistry, Physics, Earth Sciences, Engineering and related interdisciplinary fields such as Materials, Nanoscience, Chemical Engineering, Complex Systems and Biophysics.
- The work reported in the thesis must represent a significant scientific advance.
- If the thesis includes previously published material, permission to reproduce this must be gained from the respective copyright holder.
- They must have been examined and passed during the 12 months prior to nomination.
- Each thesis should include a foreword by the supervisor outlining the significance of its content.
- The theses should have a clearly defined structure including an introduction accessible to scientists not expert in that particular field.

More information about this series at http://www.springer.com/series/8790

Peter Kúš

Thin-Film Catalysts for Proton Exchange Membrane Water Electrolyzers and Unitized Regenerative Fuel Cells

Doctoral Thesis accepted by
the Charles University, Prague, Czech Republic

 Springer

Author
Dr. Peter Kúš
Department of Surface and Plasma
Science, Faculty of Mathematics
and Physics
Charles University
Prague, Czech Republic

Supervisor
Prof. Vladimír Matolín
Department of Surface and Plasma
Science, Faculty of Mathematics
and Physics
Charles University
Prague, Czech Republic

ISSN 2190-5053   ISSN 2190-5061 (electronic)
Springer Theses
ISBN 978-3-030-20861-5   ISBN 978-3-030-20859-2 (eBook)
https://doi.org/10.1007/978-3-030-20859-2

This Springer imprint is published by the registered company Springer Nature Switzerland AG
The registered company address is: Gewerbestrasse 11, 6330 Cham, Switzerland

Supervisor's Foreword

Modern society is, more than ever before, relying on a constant supply of large amount of energy. At the same time, significant public pressure is put on leading governments and industry to strengthen the shift from fossil fuels to renewable sources. The full-scale transition to renewable energy is, however, not possible without resolving certain issues. Since it is fair to expect that major part of produced energy will come from volatile sources (e.g., solar and wind), it is necessary to have a reliable and scalable way of storing and releasing it in order to meet the demands of consumers. One of the possible solutions is the implementation of hydrogen economy. The idea is that overproduced electricity would be electrochemically converted to gaseous hydrogen and oxygen via water electrolysis; generated hydrogen could be either injected into the existing natural gas pipeline, used as fuel for H_2-powered vehicles or be stored and eventually, when needed, turned back to electricity through fuel cells. The most suitable electrochemical devices for above-mentioned conversions are the proton exchange membrane water electrolyzers and fuel cells. Wider commercialization of these systems is, however, hindered by their dependence on noble metal catalysts.

In his dissertation thesis, Dr. Peter Kúš provides very deep yet comprehensive insight into the concept of hydrogen economy and approaches the problem of lowering the noble metal loading from several directions. First part of the work investigates the possibility of using magnetron sputtering for deposition of iridium onto the anode side of water electrolyzer. Although the oxygen evolution reaction which runs on the anode is exceptionally catalyst-demanding, the systematic investigation resulted in design of a unique thin-film Ir/TiC structure which performed comparable to the state-of-the-art catalysts despite utilizing just a fraction of their noble metal loading. Second part of the thesis revolves around the idea of merging the electrolyzer and fuel cell into one bifunctional system—the unitized regenerative fuel cell. This approach is convenient in applications where the complete cycle of electricity $\rightarrow H_2 \rightarrow$ electricity takes place and might lead to significant savings in catalysts, since the amount of electrodes is effectively reduced to half. Wide arsenal of analytical techniques ranging from photoelectron spectroscopy to electrochemical atomic force microscopy helped to identify and consequently address the crucial aspects which influence the performance of the bifunctional catalyst in both operational regimes. The optimized sandwich-like thin-film Ir/TiC/Pt anode catalyst yielded exceptional performance considering its very low noble metal loading.

Peter's thesis can be conceptually placed somewhere between research and application. That being said, I believe that the know-how gained from his systematic reasoning combined with "trial-and-error" approach will prove to be equally valuable as the experimental results themselves. His work might serve as a solid introductory material for students getting acquainted with the topic as well as an experimental guideline for more experienced researchers.

Prague, Czech Republic   Prof. Vladimír Matolín
March 2019

Abstract

This dissertation thesis revolves around hydrogen economy and energy-storage electrochemical systems. More specifically, it investigates the possibility of using magnetron sputtering for deposition of efficient thin-film anode catalysts with low noble metal content for proton exchange membrane water electrolyzers (PEM-WEs) and unitized regenerative fuel cells (PEM-URFCs). The motivation for this research derives from the urgent need of minimizing the price of mentioned electrochemical devices should they enter mass production.

Numerous experiments were carried out, correlating the actual in-cell performance with the varying position of thin-film catalyst within the membrane electrode assembly, with the composition of high-surface support sublayer, and with the chemical structure of the catalyst itself. The wide arsenal of analytical methods ranging from electrochemical impedance spectroscopy through scanning electron microscopy to photoelectron spectroscopy allowed us to describe complex phenomena behind different obtained efficiencies. Consequent systematic optimizations led to the design of novel PEM-WE anode thin-film iridium catalyst with thickness of just 50 nm, supported on optimized TiC-based sublayer which performed similarly to standard counterparts despite using just a fraction of their noble metal content. Moreover, the novel anode thin-film bifunctional Ir/TiC/Pt sandwich-like PEM-URFC catalyst yielded 31.15% round-trip efficiency in comparison to 40.02% given by a combination of dedicated high-loading devices.

Keywords Hydrogen economy · PEM water electrolyzer · PEM unitized regenerative fuel cell · Thin-film deposition · Low-loading catalyst

Acknowledgements

To this day, I have spent nearly seven years at the Department of Surface and Plasma Science. During this time, I have learned from and worked along very inspiring people, many of whom have become my dear and respected friends. There was always someone to help me, and I never had to ask for advice twice.

First and foremost, I would like to thank my supervisor Prof. Matolín, who introduced me to the topic, led me for the entire duration of my studies, and gave me a chance to cooperate with researchers worldwide on the international level.

Secondly, I want to thank Assoc. Prof. Matolínová for many valuable comments and recommendations regarding my work.

Special thanks to Dr. Fiala and Dr. Václavů, who knew the answer to every one of my complex questions and always pushed me in the right direction.

Next, I want to thank Dr. Ostroverkh for aiding me with MEA setup and for carrying out the fuel cell measurements.

I would also like to thank Dr. Khalakhan for his invaluable assistance during CERIC experiments and electrochemistry-related issues.

Many thanks to Dr. Kettner, Dr. Duchoň, and other colleagues from the Surface Physics Group and Nanomaterials Group who helped me countless times not only in scientific but also in personal matters.

Last but not least, a huge thanks to my parents, girlfriend, and friends outside of the academia, who kept supporting me no matter what and always fueled me with positive energy.

This work was supported by CERIC Consortium and Grant Agency of the Charles University, Projects No. 236214 and No. 1016217.

Prague, Czech Republic   Dr. Peter Kúš
July 2018

Contents

Abbreviations

AFM	Atomic force microscopy
BE	Binding energy
BSE	Backscattered electrons
CCM	Catalyst-coated membrane
EDX	Energy-dispersive X-ray spectroscopy
FC	Fuel cell
GDE	Gas diffusion electrode
GDL	Gas diffusion layer
HER	Hydrogen evolution reaction
HHV	Higher heating value
HOPG	Highly oriented pyrolytic graphite
HOR	Hydrogen oxidation reaction
IMFP	Inelastic mean free path
ITO	Indium tin oxide
LHV	Lower heating value
LOHC	Liquid organic hydrogen carrier
MEA	Membrane electrode assembly
OER	Oxygen evolution reaction
ORR	Oxygen reduction reaction
PEIS	Potentiostatic electrochemical impedance spectroscopy
PEM	Proton exchange membrane
PEM-FC	Proton exchange membrane fuel cell
PEM-URFC	Proton exchange membrane unitized regenerative fuel cell
PEM-WE	Proton exchange membrane water electrolyzer
PES	Photoelectron spectroscopy
PTFE	Polytetrafluoroethylene (i.e., Teflon)
PVD	Physical vapor deposition
RSF	Relative sensitivity factor
SE	Secondary electrons
SEM	Scanning electron microscopy

SHE	Standard hydrogen electrode
SRPES	Synchrotron radiation photoelectron spectroscopy
STP	Standard temperature and pressure
TFE	Tetrafluoroethylene
TPB	Triple-phase boundary
URFC	Unitized regenerative fuel cell
WE	Water electrolyzer
XPS	X-ray photoelectron spectroscopy

Chapter 1
Introduction

It is now year 2018 and it is evident, more than ever, that we live in Anthropocene. The epoch during which the very geology, ecosystem and climate of planet Earth are being notably altered by the human civilization. Many claim that these alterations are mostly negative. Deforestation brings certain species to the brink of extinction, overfishing and ocean pollution is responsible for drastic coral reef decline [1], even the soil samples themselves contain unambiguous traces of industrial origin such as elemental aluminium, concrete or plastics [2]. But most crucially, increased emission of greenhouse gases, mainly from burning of fossil fuels, not only pollutes the atmosphere but also causes its continual warming on the global level [3]. Taking all of the aforementioned into account, it is clear that planet Earth is now fundamentally a very different place to what it was in Holocene, merely a hundred years ago.

Even though the sentences above sound grim, the 21st century is also a time when humanity as a whole is not only widely aware of its impact on the face of the Earth but strives to take constructive action to remedy the negative impacts. Numerous states sign international treaties, such as the Kyoto Protocol or Paris Agreement [4, 5], pledging to reduce greenhouse gases emissions in order to mitigate global warming and avert further dangerous anthropogenic interference with the climate system. To do so, most of the developed countries are progressively leaning towards the utilization of renewable energy. It is not just an altruism, it is a necessity. Recent studies estimate that world reserves of coal can last for about 100 years, crude oil and natural gas for less than 30 years [6]. As such, call for diversification of energy sources and reducing the reliance on fossil fuels echoes throughout the society. Coal and natural gas power plants are being replaced by solar panels and wind turbines, automotive exhausts gases are subject to strict regulations and the presence of hybrid, electric or even hydrogen cars on the streets is no longer a fiction.

Clearly, much has been done during the last few decades, yet the ultimate goal remains, the absolute transition from fossil fuels to the decarbonized economy. In order to achieve this, novel technologies have to be implemented, capable of not only efficient power generation from renewables but also its potential conversion, storage and distribution. In another words, the generation of carbon-free energy is just half

© Springer Nature Switzerland AG 2019

P. Kúš, *Thin-Film Catalysts for Proton Exchange Membrane Water Electrolyzers and Unitized Regenerative Fuel Cells*, Springer Theses, https://doi.org/10.1007/978-3-030-20859-2_1

of the picture, the other half is having an adequate energy carrier (or energy vector) apt for substituting oil and gas. One of the possible solutions addressing these issues is the concept of so-called hydrogen economy [7].

1.1 Hydrogen Economy

The idea of using hydrogen as an ultimate alternative to the fossil fuels was proposed in the first half of the 20th century. The outlines of this concept were set out in 1923 by British scientist and Royal Society Fellow J. B. S. Haldane during his lecture, titled Daedalus or Science and the Future, at Cambridge University [8, 9]. In his talk, he stated that it is axiomatic that the exhaustion of Earth's coal and oilfields is a matter of centuries and that humanity will have to switch to intermittent but inexhaustible sources of power, the wind and the sunlight. The surplus power would be used for electrolytic decomposition of water into hydrogen and oxygen (water electrolysis). Haldane proposed liquefying the hydrogen and storing it in vast vacuum-jacketed underground reservoirs. On windless days with insufficient sunlight, the hydrogen would be converted back to electricity, either in combustion-driven dynamos or in oxidation cells (later known as fuel cells), in order to provide continual power to the grid. Hydrogen would also serve as a fuel in automotive industry and transportation in general. This Haldane's now-famous lecture and many other ideas earned him the title of "perhaps the most brilliant science popularizer of his generation" [10].

Now, nearly 100 years since its first mention it is safe to say that hydrogen economy is not just sci-fi but it is slowly yet steadily becoming reality, not too different from Haldane's vision. After all, using hydrogen as a universal energy vector offers numerous advantages. It has zero carbon footprint, provided it is created via water electrolysis powered by overproduced electricity from renewables. The raw material for this electrochemical production is water which is in high abundance. Also, hydrogen has higher specific energy (energy per unit mass) than most fuels conventionally used today (see Table 1.1).

Understandably, there are several counterarguments against the feasibility of hydrogen economy, mainly concerning storage and transportation. It is argued that gaseous hydrogen has low energy density (energy per unit volume) and that there is no dedicated transport infrastructure. These problems can however be overcome. If the relative H_2 concentration is kept within a range of 5–15% (volumetrically) it can be transported in a mixture with natural gas through existing pipelines without any significant modifications [12]. If significant improvement in energy density is the priority, hydrogen can be either compressed or liquefied [13]. Alternatively, it can be even bonded to solid hydrides or liquid organic hydrogen carriers (LOHC) [14, 15]. Moreover, LOHC in particular offers some additional benefits as it can be transported by tank trucks similarly to gasoline. In similar fashion it can also be stored in existing gas station's underground reservoirs, prior to being converted back to pure gaseous H_2, shortly before refueling a fuel cell powered vehicle [16].

Table 1.1 Specific energies and energy densities of conventionally used fuels [11]

Energy density of common fuels[a]

Fuel	Specific energy [MJ kg^{-1}]	Energy density [MJ m^{-3}]
Crude oil	42	37,000
Gasoline	46	35,000
Coal	32	42,000
Methane (gaseous)	53	38
Hydrogen (gaseous)	120	10
Hydrogen (liquid)	120	8700
Methanol	21	17,000
Ethanol	28	22,000

[a]LHV (lower heating values)

Speaking of transportation, many successful examples already exist of hydrogen-powered buses producing zero greenhouse emissions, currently operating in major European cities [17, 18]. In Lower Saxony, the first hydrogen train has completed its test run and is being set up for service [19]. Several car manufacturers are already offering or are close to releasing fuel cell cars [20–24]. Hydrogen fuel cells found their use also in scooters and industrial forklifts [25, 26]. High specific energy predetermines hydrogen for its use in aerial applications such as remote drones or even larger propeller airplanes [27, 28]. Besides transportation, fuel cell technology also penetrates other markets such as portable electronics and uninterruptible power sources [29, 30] (Fig. 1.1).

But eventually, the whole concept of hydrogen economy is only as green as the production of hydrogen itself. Considering that most of it comes from natural gas reforming, partial oxidation of oil or coal gasification, the use of such "dirty hydrogen" in green, emission-free technologies is nonsensical. The goal is to use "clean hydrogen" created via water electrolysis, powered by electricity coming from renewable sources. In the present day, unfortunately, only a tiny fraction of hydrogen produced worldwide comes from water electrolysis [31] (Fig. 1.2).

This is hopefully going to change with the take-up of next generation, cost-effective wind turbines and solar panels. Coupling them with water electrolyzers and fuel cells would achieve two goals at the same time. Firstly, coupling would help to overcome the major problem of solar and wind, the intermittency of generated power (see Fig. 1.3). The electrochemical generation of hydrogen and subsequent reconversion back to electricity would serve as a buffering mechanism [32]. Under certain circumstances, such setup proves to be much more beneficial than coupling with modern batteries [33]. Secondly, should the solar and wind power plants be installed in sufficient numbers, the surplus energy could generate enough clean hydrogen for a wide range of other above-mentioned mobile applications.

It is obvious that the backbone of hydrogen economy are efficiently working **water electrolyzers (WEs)** (i.e. devices, which use electrical power to drive elec-

Fig. 1.1 Potential application of H_2 fuel cells ranges from transportation to portable electronics

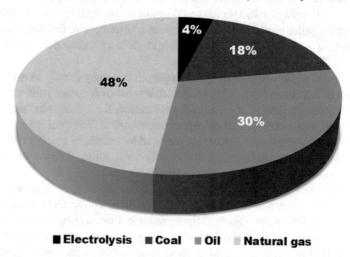

Fig. 1.2 World H_2 production breakdown

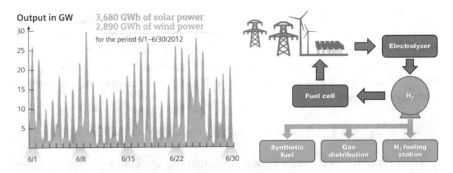

Fig. 1.3 Left—volatile nature of wind and solar output in Germany during June 2012 [34]; right—buffering mechanism for electrical grid using hydrogen as energy vector

trochemical water splitting into hydrogen and oxygen) and **fuel cells (FCs)** (i.e. devices, which generate electricity through direct electrochemical reaction of hydrogen and oxygen). Ultimately, the most elegant and optimal solution would be to have hybrid devices, the so-called **unitized regenerative fuel cells (URFCs)**, capable of operating in both "electrolyzer" and "fuel cell" regimes.

Among many different types of FCs, WEs and URFCs, the most perspective ones are arguably those utilizing the proton exchange membranes → PEM-FCs, PEM-WEs and PEM-URFCs. A major drawback of these devices, currently hindering their wider commercialization, is their dependence on noble metals which are necessary for catalyzing the redox reactions. As such, it is of utmost importance to reduce their amount without significantly deteriorating the performance and efficiency of the devices. **This dissertation addresses the above-mentioned problem and sets the goal of preparation and characterization of a novel thin-film, low-loading catalysts for PEM-WEs and PEM-URFCs**. The targets of the thesis are outlined in more detailed context later in the text after giving the necessary deeper insight into the technologies of PEM-FCs, PEM-WEs and PEM-URFCs.

Additional information regarding the concept of hydrogen economy can be found in [35–38].

1.2 Proton Exchange Membrane Fuel Cell (PEM-FC)

The fuel cell, a device which electrochemically converts chemical energy of a fuel directly into electricity[1] was first built in 1839 [39]. In this year William Grove and Christian Friedrich Schönbein carried out an experiment in which they submerged a pair of bottom-up oriented test tubes into diluted sulfuric acid (electrolyte), the test tubes were filled with oxygen and hydrogen and contained platinum electrodes

---

[1]Unlike e.g. combustion engines which convert chemical energy to heat and motion and alternatively/eventually to electricity, resulting in lower efficiency.

within. Grove and Schönbein then connected multiple of such setups (each generating approximately 1 V) into series and put the end wires into the flask of water. They observed the electrolysis of water; a proof of generated electrical current (Fig. 1.4).

The current was generated as the volume of gaseous hydrogen and oxygen in the test tubes was being reduced in 2:1 ratio, hinting to the exergonic (spontaneous) electrochemical reaction[2]:

$$2H_2 + O_2 \rightarrow 2H_2O + released\ energy \quad (+1.229\,V\ vs.\ SHE\ at\ STP) \qquad (1.1)$$

This overall reaction can be broken down into two spatially separated reactions; considering acidic electrolyte. The hydrogen oxidation reaction (HOR), taking place on anode:

$$2H_2 \rightarrow 4H^+ + 4e^- \quad (0\,V\ vs.\ SHE\ at\ STP) \qquad (1.2)$$

And oxygen reduction reaction (OER), running on cathode:

$$O_2 + 4H^+ + 4e^- \rightarrow 2H_2O \quad (+1.229\,V\ vs.\ SHE\ at\ STP) \qquad (1.3)$$

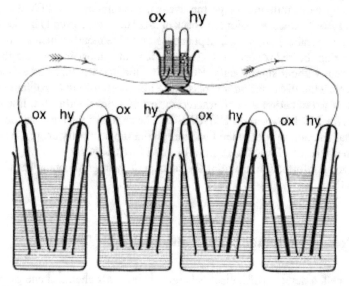

Fig. 1.4 First demonstration of hydrogen fuel cell

---

[2]Relative to SHE (standard hydrogen electrode) at STP (standard temperature and pressure) of 25 °C and 1 atm [40].

Electrons, created on anode, are flowing through the electrical circuit to the cathode, capable of doing work. H^+ ions (i.e. protons), on the other hand, are passing through the electrolyte.

Nearly 180 years since the first demonstration of the fuel cell working principle (at that time aptly called gas voltaic battery) the technology has advanced significantly. Although Grove and Schönbein proved the functionality of fuel cell using hydrogen and oxygen, in principle various other hydrogen-containing substances and oxidizing agents can serve as fuel, provided that suitable electrolyte and catalysts are used. As a result, in present day, we distinguish several types of FCs, differing in electrolyte material, catalysts, used fuel, operating temperature etc. [41].

Closest to the wider commercialization are arguably the hydrogen-fed proton exchange membrane fuel cells (PEM-FCs) [42]. As the name suggests PEM-FCs (sometimes referred to as polymer electrolyte membrane fuel cells) rely on solid state ion-conductive polymer to serve as electrolyte. Major advantages that make PEM-FCs stand out in comparison to other types of fuel cells as well as the drawbacks of this technology are summed up in Table 1.2.

The basic scheme of hydrogen fueled PEM-FC single stack is shown in Fig. 1.5.

PEM-FC is constructed in sandwich-like fashion, with an anode and a cathode on sides and with a proton exchange membrane (PEM) in the middle. Hydrogen is introduced to the anode for HOR and oxygen (either pure or simply in air) to the cathode for ORR. What differentiates it from other types of cells is the solid ion-conductive membrane, the PEM. The most widely used PEM is Nafion, developed by the E. I. DuPont Company. These membranes are produced by copolymerization of a perfluorinated vinyl ether comonomer with tetrafluoroethylene (TFE), resulting in the chemical structure [43]:

$$-[(\text{CFCF}_2)(\text{CF}_2\text{CF}_2)_m]- \atop \underset{\displaystyle \underset{\displaystyle \text{CF}_3}{|}}{\text{OCF}_2\text{CFOCF}_2\text{CF}_2\text{SO}_3\text{H}} \qquad (1.4)$$

Upon hydration of the Nafion membrane, the SO_3^- to H^+ bond considerably weakens and hydrogen ions become mobile, leading to ionic conductivity of the PEM.

Table 1.2 Pros and cons of hydrogen-fed PEM-FCs

Hydrogen-fed proton exchange membrane fuel cells	
Pros	Cons
Quick start up	Expensive Pt-based catalysts
Low temperature operation	Purity of fuel
Solid electrolyte (easy management)	
Wide load range	
High power densities	

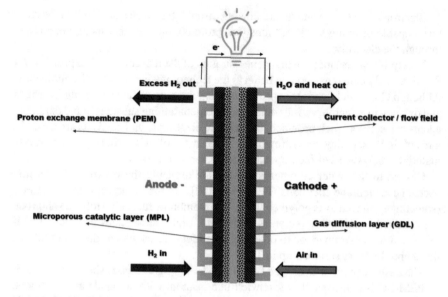

Fig. 1.5 Cross-section of hydrogen-fed PEM-FC

More details about the Nafion and other perfluorinated polymer electrolyte membranes can be found in [44].

Another key component of the PEM-FC is the gas diffusion layer (GDL). The GDL is a porous electrically conductive medium which allows the fuel to permeate from its outer side to the inner side where it meets the PEM. The most common materials used as GDLs are carbon cloth and carbon paper (Fig. 1.6).

The most crucial part of the cell is the interface between GDL and PEM, the area where the catalyst is dispersed; here the individual half-reactions take place.

Fig. 1.6 Left—carbon cloth GDL; right—carbon paper GDL [45]

This interface is called the triple phase boundary[3] (TPB). The TPB has to satisfy multiple conditions should the overall reaction run continuously. The catalyst within has to be sufficiently dispersed to maximize its specific surface and the electron and ion conductive pathways realizing transportation of e^- and H^+ have to be present. In order to achieve this, nanoparticles of catalyst are usually supported on a high-surface carbon nanopowder mixed with ionomer [46], forming the microporous catalytic layer (MPL). The mixture of catalyst, its high-surface support and ionomer is either carried on the inner side of GDL, creating gas diffusion electrode (GDE) or is directly spread over the PEM, resulting in the catalyst coated membrane (CCM). Regardless of which variant is used, the supported catalyst ends up being in between GDL and PEM (see Fig. 1.5).

Since the PEM requires hydration to be functional and one of the overall reaction's products is H_2O, it is obvious that water management of the cell is of great importance [47, 48]. Therefore, a certain amount of polytetrafluorethylen (PTFE i.e. Teflon) is often introduced to the TPB and/or on the surface of GDL.

As stated in the Table 1.2, the main drawback of PEM-FC technology, currently hindering its wider commercialization, is the reliance on noble metal catalysts; more specifically platinum. Decades of research and development have proven that from the electrochemical point of view platinum is the optimal choice for both HOR and ORR [49]. It is able to adsorb the reactants strongly enough to facilitate a reaction but not so much that the catalytic site becomes blocked by the products. In addition, it is stable enough to withstand harsh operational conditions of the cell such as pH between 2 and 3, temperature around 80 °C and high applied voltage. Unfortunately, Pt belongs to the least abundant elements in the Earth's crust which is reflected in its price [50, 51]. Based on several studies, the cost of electrocatalyst represents roughly 30–50% of the PEM-FC stack's price (depending on its size and output power) [52]. Although significant Pt loading reduction has been achieved over the last couple of years[4] the amount of noble metal still needs to be decreased to make PEM-FCs economically viable.[5] To do so, various approaches are being proposed, including further enhancement of Pt dispersion on high-surface supports [54, 55], alloyzation of Pt with non-noble metals and creating core-shell structures [56, 57] or incorporating Pt nanoparticles within the transition metal oxides such as CeO_2 [58].

For more information about PEM-FCs, refer to [59, 60].

---

[3]This term has its origins in era of early fuel cells which utilized solid electrodes, liquid electrolyte and gaseous fuel, hence three phases are entering the reaction.

[4]Thanks to carbon black support, typical noble metal loading for cathode and anode has been lowered from several mg cm^{-2} down to 0.5–0.3 mg cm^{-2} [42].

[5]DOE target for 2020 is set to 14 \$/kW [53].

1.3 Proton Exchange Membrane Water Electrolyzer (PEM-WE)

The roots of first experiments regarding water electrolysis go back to 1800, when William Nicholson and Anthony Carlisle used voltaic pile (invented by Alessandro Volta in the same year) for the splitting of water into hydrogen and oxygen [61]. Electrochemically speaking, water electrolysis is an endergonic (nonspontaneous) reaction, i.e. energy is absorbed:

$$2H_2O + absorbed\ energy \rightarrow 2H_2 + O_2 \quad (+1.229\,V \text{ vs. SHE at STP}) \quad (1.5)$$

Considering acidic electrolyte, the overall redox reaction consists of two half-reactions. Oxygen evolution reaction (OER) on anode (oxidation):

$$2H_2O \rightarrow O_2 + 4H^+ + 4e^- \quad (+1.229\,V \text{ vs. SHE at STP}) \quad (1.6)$$

And hydrogen evolution reaction (HER) on cathode (reduction):

$$2H^+ + 2e^- \rightarrow H_2 \quad (0\,V \text{ vs. SHE at STP}) \quad (1.7)$$

Although, the most established and well matured method for electrolytic hydrogen production at a commercial level is alkaline water electrolysis [62], other modern technologies are starting to take over; most notably proton exchange membrane water electrolyzers (PEM-WEs) [63]. Advantages and disadvantages of PEM-WEs in comparison to alkaline and other types of electrolyzers are listed in Table 1.3.

Clearly, the pros and cons of PEM-WE technology somehow mimic the attributes of PEM-FCs. This is to be expected, after all PEM-WE and PEM-FCs share many similarities. However, there are also some key differences. Simplified scheme of PEM-WE single stack is shown in Fig. 1.7.

The cathode side of the PEM-WE cell including the membrane does not differ much from the anode of the PEM-FC. Sure, the H_2 flow goes, in case of PEM-WE, outwards from the PEM but the GDL structure and Pt catalyst are basically the same.

Table 1.3 Pros and cons of PEM-WEs

Proton exchange membrane water electrolyzers	
Pros	Cons
Dynamic operation	Expensive Pt and Ir based catalysts
Low temperature	Purity of water
Solid electrolyte (easy management)	High anodic potentials
High current densities	Possible low durability due to corrosion
High gas purity	

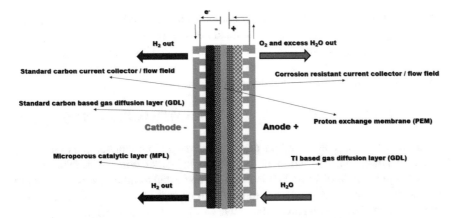

Fig. 1.7 Cross-section of PEM-WE

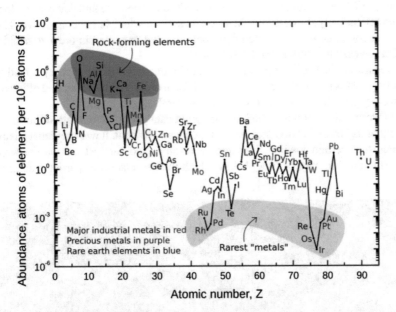

Fig. 1.8 Abundance of selected elements in the Earth's crust [50]

On the other hand, the anode side of PEM-WE does not resemble the cathode of the PEM-FC at all. The most apparent difference is the absence of platinum. It turns out, Pt is not the optimal catalyst for OER. The state-of-the-art catalyst is Ir, IrO_2 respectively [64]. Unfortunately, Ir is even less abundant in the Earth's crust than Pt, about an order of magnitude less to be precise [50] (Fig. 1.8).

Evidently, this represents a great obstacle to potential mass production. In addition, reducing the amount of Ir on anode of PEM-WE proves to be more challenging than lowering the loading of Pt on cathode of PEM-WE or on both electrodes of

PEM-FC. The OER's kinetics is slower compared to HER and as such more catalyst demanding [65]. What complicates the matter even further is the high anodic potential, present during the operation of the PEM-WE cell. This potential, typically in a range of 1.5–2 V versus SHE during operation, causes rapid carbon oxidation [40, 66], rendering carbon nanostructures inapplicable:

$$C + 2H_2O \rightarrow CO_2 + 4H^+ + 4e^- \quad (+0.207\,\text{V vs. SHE at STP}) \quad\quad (1.8)$$

In other words, neither carbon paper nor carbon cloth can serve as GDL and carbon black cannot be used as high-surface catalyst support.[6] A notable part of the R&D in the field of PEM-WE is therefore dedicated to finding a material apt to substitute carbon. It obviously has to be electrochemically inert to withstand high anodic potential and sufficiently conductive to allow electron transfer.

Various Ti based meshes and felts can serve as GDL (see Fig. 1.9). Even though during PEM-WE operational conditions Ti does not corrode and decay as C does, it electrochemically forms a TiO_2 layer on its surface which is nonconductive. Therefore, Ti components have to be additionally treated or coated in order to avoid this oxidation (usually with a thin layer of gold).

Regarding the OER catalyst itself, it can be unsupported, i.e. simply mixed with ionomer and spread on the membrane or GDL. In order to achieve desired efficiencies however, the loading has to be very high (several mg cm^{-2}) [63].

Lowering the amount of such unsupported catalyst or attempting to increase the extent of its dispersion does not lead to much improvement. It was previously demonstrated that certain loading threshold of unsupported catalyst can be determined (0.5 mg cm^{-2} in case of IrO_2) below which the percolation of particles and hence

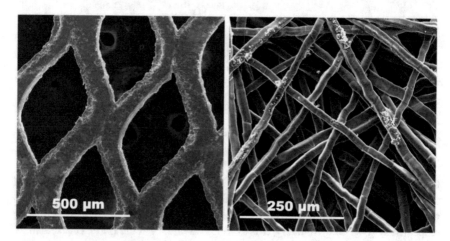

Fig. 1.9 Left—Ti mesh anode GDL; right—Ti felt anode GDL

---

[6]It should be emphasized that usage of these carbon nanostructures was a pivotal factor, allowing for considerable Pt reduction in PEM-FCs without their significant efficiency deterioration.

the electron conductivity is insufficient and performance considerably worsens [67]. In order to further cut down the amount of iridium, the usage of suitable catalyst support is reportedly inevitable [68]. Generally speaking, metal oxides exhibit good corrosion resistance, their electric conductivity is poor though. Nonetheless, materials such as TiO_2 [69], ITO [70], SnO_2 [71], or tantalum oxide [72] were successfully used as anode catalyst supports. Systems with catalyst dispersed on metal oxide yielded better in-cell PEM-WE performance in comparison to pure unsupported counterparts with the same noble metal loading. The conclusion is that to a certain extent conductivity of well dispersed catalyst alone is sufficient enough to ensure low ohmic resistance of catalyst/metal oxide layer [69]. This however ceases to apply once the catalyst loading is significantly decreased and conductive percolation path disintegrates.

To sum up, the structure of PEM-WE is quite similar to that of PEM-FC, yet there are some crucial differences, due to the high operational potential of OER. Increasing the efficiency and lowering the catalyst loading in PEM-WEs is therefore mainly a question of optimizing the anode side of the cell.

More details about PEM-WEs are provided in [73, 74].

1.4 Proton Exchange Membrane Unitized Regenerative Fuel Cell (PEM-URFC)

The concept of Hydrogen economy is built around the idea of repetitive conversion of surplus energy to hydrogen and eventually back to electricity. Dedicated electrochemical devices for individual processes, PEM-FC and PEM-WE, were thoroughly described in previous sections. The prices of noble metal catalysts and bipolar plates have been identified as a main obstacle should these technologies enter mass production (see Fig. 1.10) [53].

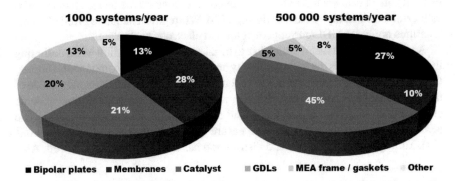

Fig. 1.10 Breakdown of PEM-FC stack cost at 1000 and 500,000 systems per year (price ratios of PEM-WE components are expected to be quite similar)

Merging PEM-FC and PEM-WE cells into one bifunctional device would significantly reduce the costs[7] in applications where these two systems work in close synergy, predominantly in the balancing of renewable energy-powered electrical grid. Such device capable of both regimes of operation, water splitting as well as power generating, is called unitized regenerative fuel cell (URFC).

Even though the first concepts of PEM-URFC were put forward in the early sixties of 20th century, the development and construction of a universal device which would have equal performance and round-trip efficiency as coupled state-of the-art PEM-WE and PEM-FC remains a challenge [75].

By definition PEM-URFC needs to execute four different redox reactions, namely HOR, ORR, HER and OER. Three of these reactions can be effectively catalyzed by Pt (HOR, ORR and HER), OER however requires Ir.[8] The question therefore arises how to arrange the cell in order for it to function both ways, or in another words, which PEM-FC reaction should run on the same side as OER. Principally, there are two possibilities: **A** having HOR/HER on one side, ORR/OER on another (H_2 and O_2 electrode configuration); or **B** having HOR/OER on one side, HER/ORR on the other (oxidation and reduction electrode configuration). For more detail, refer to Fig. 1.11. Configuration **A** keeps the same gases on same electrodes during both regimes. The obvious advantage of this concept is much simpler gas management. On the other hand, ORR and OER, both of which are in terms of kinetics slower reactions of individual regimes (so to say bottlenecks of the overall reaction's rate), are kept on the same side of the cell. As such, additional high demands are put on the stability and reliability of bifunctional Ir–Pt based catalyst, as its potential defectiveness would majorly affect the URFC functionality. Configuration **A** seems to be the mainstream of current R&D [76–78]. Configuration **B** pairs OER with HOR; a slow reaction with relatively fast one (in similar fashion HER and ORR are paired on the other side of the cell). The electrodes do not change their redox role, anode stays anode and cathode stays cathode. The downside of this variant is the need for gas line purging prior to the PEM-WE/PEM-FC regime change [75, 79, 80].

Needless to say, whichever reaction gets to run on the same side as OER (HOR or ORR), its Pt catalyst has to be supported on the same corrosion resistant material as Ir catalyst, since high potential during PEM-WE regime cannot be avoided. Same measures apply for GDL/current collector and electrode back plate.

Another important factor which has to be taken into consideration when designing the PEM-URFC cell is the hydrophobicity and hydrophilicity of components forming the MEA. Certain compromise has to be found particularly on the bifunctional side of the cell, where OER favors hydrophilicity, yet ORR or HOR run better in hydrophobic environment. An ideal ratio of PTFE and Nafion within the microporous catalytic layer has to be established in order to ensure sufficient efficiency of both reactions [81]. Tuning of the pore's size and distribution has a significant effect as well. Also

---

[7]Not only the costs of catalysts and bipolar plates but also all other parts of the cell since their number basically gets reduced to half.

[8]Now for simplicity considering state-of-the-art, established catalysts (Pt and Ir); not novel, experimental ones.

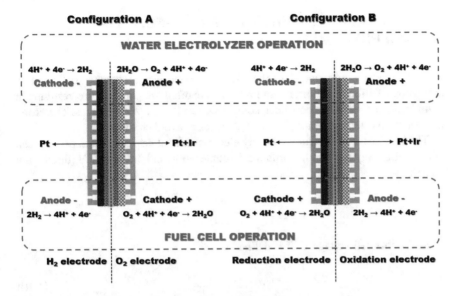

Fig. 1.11 Different PEM-URFC electrode configurations: left—H_2 and O_2 electrode configuration; right—oxidation and reduction electrode configuration

the membrane itself has to be subject to compromise; dedicated PEM-FC membranes operate under relative humidity of about 80%, the PEM-WE membranes are however in full contact with water [48].

Considering the bifunctional catalyst itself (regardless if in conf. **A** or conf. **B**), there is no unanimous opinion on what is the ideal distribution and ratio of Ir and Pt within the MPL. Numerous research groups experimented with different setups. To name a few, Pt sprayed on GDL and IrO_2 on top of Pt or Pt on membrane and IrO_2 on GDL [82]; multilayered Pt–IrO_2 structures, segmented chessboard-like structures and one-layer composite Pt+IrO_2 structures [83], alloyed Pt–Ir layer and simple nanoparticle mixtures [76]. Different studies often seem to contradict each other in terms of ideal distribution; regarding the ratio though the straightforward rule of thumb is more Pt to promote HOR/ORR of PEM-FC regime, more Ir to promote OER of PEM-WE regime. Additional in-depth insight into various tested ratios and configurations can be found elsewhere [64, 75, 84].

To sum up, PEM-URFC is still clearly somehow immature technology in comparison to PEM-FC and PEM-WE and intensive R&D has to continue in order to broaden the knowledge base. Due to the high price and many unresolved technological challenges, PEM-URFCs are currently limited mainly to aerospace and military applications.

1.5 Performance and Efficiency of PEM-FC, PEM-WE and PEM-URFC

Talking about the efficiencies of the aforementioned electrochemical cells can be ambiguous. Different definitions and wording are often used in literature, resulting in undesirable incomprehension when comparing results of several sources. Therefore, it is sound to briefly go through essential basics to avoid confusion.

Thermodynamic efficiency η of any electrochemical device is the ratio between Gibbs free energy ΔG (the maximum extractible work) and enthalpy ΔH (the heating value):

$$\eta = \frac{\Delta G}{\Delta H} \tag{1.9}$$

Reversible cell voltage E^{rev} is defined as:

$$E^{rev} = -\frac{\Delta G^0}{nF} \tag{1.10}$$

where n is the amount of transferred electrons and $F = 96487\,C\,mol^{-1}$ is the Faraday's constant. In case of water formation (in liquid form) the $\Delta G^0 = -237.13\,kJ\,mol^{-1}$ (at STP) and $n = 2$. Hence, the reversible voltage for hydrogen fuel cell at standard conditions is $E^{rev} = 1.229\,V$. This is the maximum voltage which can in theory be obtained from the electrochemical reaction of H_2 and O_2.

The assumption that all of hydrogen's chemical energy (i.e. its heating value) was converted to electric energy (although this is not possible) would lead to the thermoneutral voltage:

$$E^{th} = -\frac{\Delta H^0}{nF} \tag{1.11}$$

Taking the higher heating value (HHV) of hydrogen into consideration $\Delta H^0_{HHV} = -285.83\,kJ\,mol^{-1}$ (at STP), the thermoneutral voltage for hydrogen cell at standard conditions is $E^{th} = 1.481\,V$.

Putting Eqs. (1.9)–(1.11) together, the maximum thermodynamic efficiency of hydrogen fuel cell at standard conditions can be calculated as:

$$\eta^{max} = \frac{\Delta G^0}{\Delta H^0} = \frac{E^{rev}}{E^{th}} = 83.1\% \tag{1.12}$$

Since hydrogen fuel cells practically never operate at room temperature[9] and at reversible potential, it is convenient to be able to calculate efficiency at any given cell voltage and temperature.

---

[9]Both ΔG and ΔH are temperature dependent.

If the inputs are the instantaneous voltage of fuel cell E^{FC} and enthalpy at the given temperature ΔH^T, Eq. (1.9) takes the form of:

$$\eta^{FC} = \frac{nFE^{FC}}{\Delta H^T} \tag{1.13}$$

Considering water splitting at voltage E^{WE} instead of water formation, the efficiency η^{WE} is given by:

$$\eta^{WE} = \frac{\Delta H^T}{nFE^{WE}} \tag{1.14}$$

To demonstrate a practical example of efficiency calculation, let us imagine PEM-WE operating at 80 °C the IV curve[10] of which is plotted in Fig. 1.12.

To calculate the η^{WE} at current density of $500\,\mathrm{mA\,cm^{-2}}$, we need the corresponding operational voltage (in this case $E^{WE} = 1.7\,\mathrm{V}$) and the high heating value of hydrogen at 80 °C, which can be found in chemical tables (in this case $\left(\Delta H^{80}_{HHV} = -284.04\,\mathrm{kJ\,mol^{-1}}\right)$). Using relation (1.14), the calculated PEM-WE efficiency is $\eta^{WE} = 86.58\%$.

Now let us consider the PEM-FC operating at 80 °C and performing as plotted in Fig. 1.13. Analogically, to obtain the efficiency η^{FC} at current density of $500\,\mathrm{mA\,cm^{-2}}$, the necessary inputs are the operational voltage $\left(E^{FC} = 0.6\,\mathrm{V}\right)$ and the HHV of hydrogen at 80 °C $\left(\Delta H^{80}_{HHV} = -284.04\,\mathrm{kJ\,mol^{-1}}\right)$. Using relation (1.13), the calculated PEM-FC efficiency is $\eta^{FC} = 40.76\%$.

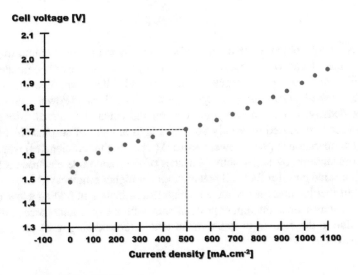

Fig. 1.12 Example of PEM-WE IV curve

---

[10] Also called polarization curve.

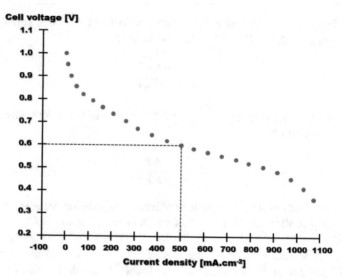

Fig. 1.13 Example of PEM-FC IV curve

Overall round-trip efficiency of coupled PEM-WE and PEM-FC is $\eta^{RT} = \eta^{WE}\eta^{FC} = 35.29\%$. If the water electrolyzer and fuel cell operate at the same temperature,[11] the HHV is the same for both partial efficiencies, hence the round-trip efficiency can be calculated simply as:

$$\eta^{RT} = \frac{E^{FC}}{E^{WE}} \tag{1.15}$$

It is now obvious that in order to increase the round-trip efficiency, the PEM-WE would have to operate at lower voltages and PEM-FC at higher voltages. This would however inevitably lead to the lower input and output power densities (Power density = Current density × Cell voltage) as seen on individual IV curves. Since water electrolyzer is supposed to readily convert a high amount of surplus electricity to hydrogen, operation at lower power densities (albeit higher efficiencies) would have to be compensated by larger active surfaces of the membrane electrode assembly (MEA[12]); same goes for fuel cell power output at higher efficiency.

This further justifies the researcher's high determination in finding a low-loading effective catalyst since the higher price of cells with a larger active area is predominantly due to the amount of noble metal within (considering mass production of these devices).

---

[11]Or unitized regenerative fuel cell operates at the same temperature in both regimes.

[12]PEM, catalyst layers (anode and cathode), and GDL together form the MEA.

1.6 Thesis Motivation and Targets

The previous chapters comprehensively introduced the PEM-FCs, PEM-WEs and PEM-URFCs. These systems are undoubtedly highly relevant in the context of modern self-sustainable, renewables-oriented society. The R&D still faces numerous challenges, which have to be solved prior to the mass market entry of mentioned PEM technologies. Noble metal catalyst reduction tops the list of most crucial issues.

In our previous research at the Department of Surface and Plasma Science, the studies were focused primarily on fuel cells. We obtained several interesting results concerning novel Pt low-loading materials for the anode and cathode side of PEM-FC and published them in high impact factor journals [54, 55, 58, 85, 86]. Recently we have turned our attention to PEM-WEs and PEM-URFCs. The ultimate goal remains the same: to prepare, characterize and test novel nanostructured low-loading catalytically active materials. In this work we will be focusing on the more catalyst-demanding anode side of the cell. The novelty of our approach lies in utilizing thin-film magnetron sputtering[13] for catalyst deposition which offers numerous advantages over more conventional methods of preparation; such as: industrial cost-effectiveness and scalability, excellent homogeneity of the sputtered material, ability to controllably and reproducibly deposit very low catalyst loadings and possibility to prospectively prepare not only pure metals but alloys and compounds as well, utilizing co-sputtering and reactive sputtering respectively.

The objectives of the thesis have been set as follows:

1. Construct the PEM-WE/PEM-URFC experimental cell and testing station.
2. Investigate the possibilities of using thin-film low-loading catalysts (less than 0.3 mg cm^{-2}) for the anode side of the PEM-WE cell (the oxygen evolution electrode).
3. Investigate the possibilities of using thin-film low-loading catalyst in the PEM-URFC cell.

References

1. Zaneveld JR, Burkepile DE, Shantz AA, Pritchard CE, McMinds R, Payet JP, Welsh R, Correa AMS, Lemoine NP, Rosales S, Fuchs C, Maynard JA, Thurber RV (2016) Overfishing and nutrient pollution interact with temperature to disrupt coral reefs down to microbial scales. Nat Commun 7:11833. https://doi.org/10.1038/ncomms11833
2. Waters CN, Zalasiewicz J, Summerhayes C, Barnosky AD, Poirier C, Gauszka A, Cearreta A, Edgeworth M, Ellis EC, Ellis M, Jeandel C, Leinfelder R, McNeill JR, de Richter D, Steffen W, Syvitski J, Vidas D, Wagreich M, Williams M, Zhisheng A, Grinevald J, Odada E, Oreskes N, Wolfe AP (2016) The Anthropocene is functionally and stratigraphically distinct from the Holocene. Science 351. https://doi.org/10.1126/science.aad2622
3. Le Treut H, Somerville R, Cubasch U, Ding Y, Mauritzen C, Mokssit A, Peterson T, Prather M (2007) Historical overview of climate change science coordinating. In: Climate change

---

[13]Sputtering is thoroughly described in the Experimental section.

2007—the physical science basis: working group I contribution to the fourth assessment report of the Intergovernmental Panel on Climate Change

4. U. Nations (1998) Kyoto protocol to the United Nations framework. Rev Eur Community Int Environ Law. https://doi.org/10.1111/1467-9388.00150
5. U. Nations (2015) Paris Agreement. Conf. Parties Its Twenty-First Sess. https://undocs.org/FCCC/CP/2015/L.9/Rev.1
6. Shafiee S, Topal E (2009) When will fossil fuel reserves be diminished? Energy Policy 37:181–189. https://doi.org/10.1016/j.enpol.2008.08.016
7. Armaroli N, Balzani V (2011) The hydrogen issue. ChemSusChem 4:21–36. https://doi.org/10.1002/cssc.201000182
8. Haldane JBS (1923) Daedalus, or, science and the future, A Pap. Read to Heretics, Cambridge. http://bactra.org/Daedalus.html. Accessed 21 June 2018
9. Roberts P (2005) The end of oil: on the edge of a perilous new world. Houghton Mifflin
10. Haldane JBS, Dronamraju KR (2009) What I require from life: writings on science and life from J.B.S. Haldane. Oxford University Press
11. Pahwa GK, Pahwa PK (2014) Hydrogen economy. The Energy and Resources Institute (TERI)
12. Melaina MW, Antonia O, Penev M (2013) Blending hydrogen into natural gas pipeline networks: a review of key issues. National Renewable Energy Laboratory. https://doi.org/10.2172/1068610
13. Berry GD, Aceves SM (1998) Onboard storage alternatives for hydrogen vehicles. Energy Fuels 12:49–55. https://doi.org/10.1021/ef9700947
14. Orimo SI, Nakamori Y, Eliseo JR, Züttel A, Jensen CM (2007) Complex hydrides for hydrogen storage. Chem Rev 107:4111–4132. https://doi.org/10.1021/cr0501846
15. He T, Pei Q, Chen P (2015) Liquid organic hydrogen carriers. J Energy Chem 24:587–594. https://doi.org/10.1016/j.jechem.2015.08.007
16. Preuster P, Papp C, Wasserscheid P (2017) Liquid Organic Hydrogen Carriers (LOHCs): toward a hydrogen-free hydrogen economy. Acc Chem Res 50:74–85. https://doi.org/10.1021/acs.accounts.6b00474
17. Gangi J, Curtin S (2015) Fuel cell technologies market report 2014. U.S. Dep. Energy
18. Stempien JP, Chan SH (2017) Comparative study of fuel cell, battery and hybrid buses for renewable energy constrained areas. J Power Sources 340:347–355. https://doi.org/10.1016/j.jpowsour.2016.11.089
19. Molloy M (2017) The world's first zero-emissions hydrogen train is coming. The Telegraph. http://www.telegraph.co.uk/technology/2017/03/22/worlds-first-zero-emissions-hydrogen-train-coming/. Accessed 23 Aug 2017
20. Eberle U, Müller B, von Helmolt R (2012) Fuel cell electric vehicles and hydrogen infrastructure: status 2012. Energy Environ Sci 5. https://doi.org/10.1039/c2ee22596d
21. Yoshida T, Kojima K (2015) Toyota MIRAI fuel cell vehicle and progress toward a future hydrogen society. Interface Mag 24:45–49. https://doi.org/10.1149/2.F03152if
22. Wiseman E (2018) Hyundai Nexo review—driving the next-generation hydrogen fuel cell SUV in Britain for the first time. The Telegraph. https://www.telegraph.co.uk/cars/hyundai/hyundai-nexo-review-driving-next-generation-hydrogen-fuel-cell/. Accessed 25 June 2018
23. Under the microscope: Mercedes-Benz GLC F-CELL: the fuel cell gets a plug. Daimler Global Media Site. http://media.daimler.com/marsMediaSite/en/instance/ko/Under-the-microscope-Mercedes-Benz-GLC-F-CELL-The-fuel-cell-gets-a-plug.xhtml?oid=11111320. Accessed 23 Aug 2017
24. Clarity fuel cell—environmentally-conscious vehicles. Honda Global Media Site. https://automobiles.honda.com/clarity. Accessed 23 Aug 2017
25. Elgowainy A, Gaines L, Wang M (2009) Fuel-cycle analysis of early market applications of fuel cells: forklift propulsion systems and distributed power generation. Int J Hydrogen Energy 34:3557–3570. https://doi.org/10.1016/j.ijhydene.2009.02.075
26. Hwang JJ (2012) Review on development and demonstration of hydrogen fuel cell scooters. Renew Sustain Energy Rev 16:3802–3815. https://doi.org/10.1016/j.rser.2012.03.036

27. Rees M (2015) Horizon unveils hydrogen-powered multirotor UAV. Unmanned Systems Technology. http://www.unmannedsystemstechnology.com/2015/05/horizon-unveils-hydrogen-powered-multirotor-uav/. Accessed 24 Aug 2017
28. Khandelwal B, Karakurt A, Sekaran PR, Sethi V, Singh R (2013) Hydrogen powered aircraft: the future of air transport. Prog Aerosp Sci 60:45–59. https://doi.org/10.1016/j.paerosci.2012.12.002
29. Kundu A, Jang JH, Gil JH, Jung CR, Lee HR, Kim SH, Ku B, Oh YS (2007) Micro fuel cells: current development and applications. J Power Sources 170:67–78. https://doi.org/10.1016/j.jpowsour.2007.03.066
30. Choi W, Howze JW, Enjeti P (2006) Fuel-cell powered uninterruptible power supply systems: design considerations. J Power Sources 157:311–317. https://doi.org/10.1016/j.jpowsour.2005.07.058
31. Abbasi T, Abbasi SA (2011) "Renewable" hydrogen: prospects and challenges. Renew Sustain Energy Rev 15:3034–3040. https://doi.org/10.1016/j.rser.2011.02.026
32. Gahleitner G (2012) Hydrogen from renewable electricity: an international review of power-to-gas pilot plants for stationary applications. Int J Hydrogen Energy 38:2039–2061. https://doi.org/10.1016/j.ijhydene.2012.12.010
33. Pellow MA, Emmott CJM, Barnhart CJ, Benson SM (2015) Hydrogen or batteries for grid storage? A net energy analysis. Energy Environ Sci 8:1938–1952. https://doi.org/10.1039/C4EE04041D
34. Ruth C (2014) A better forecast for renewable energy generation. Siemens Global Media Site. https://www.siemens.com/innovation/en/home/pictures-of-the-future/energy-and-efficiency/sustainable-power-generation-neural-networks.html. Accessed 18 Oct 2017
35. Subramani V, Basile A, Veziroğlu TN (2015) Compendium of hydrogen energy. Hydrogen production and purification, vol 1. Woodhead Publishing
36. Gupta R, Basile A, Veziroğlu TN (2015) Compendium of hydrogen energy. Hydrogen storage, distribution and infrastructure, vol 2. Woodhead Publishing
37. Barbir F, Basile A, Veziroğlu TN (2015) Compendium of hydrogen energy. Hydrogen energy conversion, vol 3. Woodhead Publishing
38. Ball M, Basile A, Veziroğlu TN (2015) Compendium of hydrogen energy. Hydrogen use, safety and the hydrogen economy, vol 4. Woodhead Publishing
39. Hoogers G (2003) Fuel cell technology handbook. CRC Press
40. Bard AJ, Parsons R, Jordan J (1985) Standard potentials in aqueous solution. International Union of Pure and Applied Chemistry, M. Dekker, New York
41. Vielstich W, Lamm A, Gasteiger HA, Yokokawa H (2003) Handbook of fuel cells: fundamentals, technology, and applications. Wiley
42. Wang Y, Chen KS, Mishler J, Cho SC, Adroher XC (2011) A review of polymer electrolyte membrane fuel cells: technology, applications, and needs on fundamental research. Appl Energy 88:981–1007. https://doi.org/10.1016/j.apenergy.2010.09.030
43. Mauritz KA, Moore RB (2004) State of understanding of Nafion. Chem Rev 104:4535–4585. https://doi.org/10.1021/cr0207123
44. Rikukawa M, Sanui K (2000) Proton-conducting polymer electrolyte membranes based on hydrocarbon polymers. Prog Polym Sci 25:1463–1502. https://doi.org/10.1016/S0079-6700(00)00032-0
45. What is the difference between carbon paper and carbon cloth based Gas Diffusion Layers (GDL)?, Fuel Cells Etc (2013). http://fuelcellsetc.com/2013/03/comparing-gas-diffusion-layers-gdl/. Accessed 30 Aug 2017
46. Antolini E (2009) Carbon supports for low-temperature fuel cell catalysts. Appl Catal B Environ 88:1–24. https://doi.org/10.1016/j.apcatb.2008.09.030
47. Li H, Tang Y, Wang Z, Shi Z, Wu S, Song D, Zhang J, Fatih K, Zhang J, Wang H, Liu Z, Abouatallah R, Mazza A (2008) A review of water flooding issues in the proton exchange membrane fuel cell. J Power Sources 178:103–117. https://doi.org/10.1016/j.jpowsour.2007.12.068

48. Zhang J, Tang Y, Song C, Xia Z, Li H, Wang H, Zhang J (2008) PEM fuel cell relative humidity (RH) and its effect on performance at high temperatures. Electrochim Acta 53:5315–5321. https://doi.org/10.1016/j.electacta.2008.02.074
49. Holton OT, Stevenson JW (2013) The role of platinum in proton exchange membrane fuel cells. Platin Met Rev 57:259–271. https://doi.org/10.1595/147106713X671222
50. Parry SJ (1984) Abundance and distribution of palladium, platinum, iridium and gold in some oxide minerals. Chem Geol 43:115–125. https://doi.org/10.1016/0009-2541(84)90142-6
51. Platinum group metals: price charts. Johnson Matthey. http://www.platinum.matthey.com/prices/price-charts. Accessed 1 Jan 2017
52. DOE hydrogen and fuel cells program: 2008 annual merit review proceedings. Energy.Gov. https://www.hydrogen.energy.gov/annual_review08_proceedings.html. Accessed 31 Aug 2017
53. Marcinkoski J, Spendelow J, Wilson A, Papageorgopoulos D (2015) DOE hydrogen and fuel cells program record: fuel cell system cost—2015. Energy.Gov. https://www.hydrogen.energy.gov/pdfs/15015_fuel_cell_system_cost_2015.pdf. Accessed 19 Oct 2017
54. Dvořák F, Camellone MF, Tovt A, Tran N, Negreiros FR, Vorokhta M, Skála T, Matolínová I, Mysliveček J, Matolín V, Fabris S (2016) Creating single-atom Pt-ceria catalysts by surface step decoration. Nat Commun 7:1–8. https://doi.org/10.1038/ncomms10801
55. Bruix A, Lykhach Y, Matolínová I, Neitzel A, Skála T, Tsud N, Vorokhta M, Stetsovych V, Ševčíková K, Mysliveček J, Fiala R, Václavů M, Prince KC, Bruyere S, Potin V, Illas F, Matolín V, Libuda J, Neyman KM (2014) Maximum noble-metal efficiency in catalytic materials: atomically dispersed surface platinum. Angew Chemie Int Ed 53:10525–10530. https://doi.org/10.1002/anie.201402342
56. Vorokhta M, Khalakhan I, Václavů M, Kovács G, Kozlov SM, Kúš P, Skála T, Tsud N, Lavková J, Potin V, Matolínová I, Neyman KM, Matolín V (2016) Surface composition of magnetron sputtered Pt-Co thin film catalyst for proton exchange membrane fuel cells. Appl Surf Sci 365:245–251. https://doi.org/10.1016/j.apsusc.2016.01.004
57. Gasteiger HA, Kocha SS, Sompalli B, Wagner FT (2005) Activity benchmarks and requirements for Pt, Pt-alloy, and non-Pt oxygen reduction catalysts for PEMFCs. Appl Catal B Environ 56:9–35. https://doi.org/10.1016/j.apcatb.2004.06.021
58. Fiala R, Figueroba A, Bruix A, Vaclavu M, Rednyk A, Khalakhan I, Vorokhta M, Lavkova J, Illas F, Potin V, Matolinova I, Neyman KM, Matolin V (2016) High efficiency of Pt^{2+}-CeO_2 novel thin film catalyst as anode for proton exchange membrane fuel cells. Appl Catal B Environ 197:262–270. https://doi.org/10.1016/j.apcatb.2016.02.036
59. Barbir F (2013) PEM fuel cells: theory and practice. Academic Press
60. Qi Z (2017) Proton exchange membrane fuel cells. CRC Press
61. Rieger PH (1994) Chapter: Electrolysis, electrochemistry. Springer, pp 71–426. https://doi.org/10.1007/978-94-011-0691-7_7
62. Ursua A, Gandia LM, Sanchis P (2012) Hydrogen production from water electrolysis: current status and future trends. Proc IEEE 100:410–426. https://doi.org/10.1109/JPROC.2011.2156750
63. Carmo M, Fritz DL, Mergel J, Stolten D (2013) A comprehensive review on PEM water electrolysis. Int J Hydrogen Energy 38:4901–4934. https://doi.org/10.1016/j.ijhydene.2013.01.151
64. Antolini E (2014) Iridium as catalyst and cocatalyst for oxygen evolution/reduction in acidic polymer electrolyte membrane electrolyzers and fuel cells. ACS Catal 4:1426–1440. https://doi.org/10.1021/cs4011875
65. Damjanovic A, Dey A, Bockris JO (1966) Electrode kinetics of oxygen evolution and dissolution on Rh, Ir, and Pt-Rh alloy electrodes. J Electrochem Soc 113:739. https://doi.org/10.1149/1.2424104
66. Kim J, Lee J, Tak Y (2009) Relationship between carbon corrosion and positive electrode potential in a proton-exchange membrane fuel cell during start/stop operation. J Power Sources 192:674–678. https://doi.org/10.1016/j.jpowsour.2009.03.039

67. Rozain C, Mayousse E, Guillet N, Millet P (2016) Influence of iridium oxide loadings on the performance of PEM water electrolysis cells: Part I—Pure IrO_2-based anodes. Appl Catal B Environ 182:153–160. https://doi.org/10.1016/j.apcatb.2015.09.013

68. Rozain C, Mayousse E, Guillet N, Millet P (2016) Influence of iridium oxide loadings on the performance of PEM water electrolysis cells: Part II—Advanced oxygen electrodes. Appl Catal B Environ 182:123–131. https://doi.org/10.1016/j.apcatb.2015.09.011

69. Mazúr P, Polonský J, Paidar M, Bouzek K (2012) Non-conductive TiO_2 as the anode catalyst support for PEM water electrolysis. Int J Hydrogen Energy 37:12081–12088. https://doi.org/10.1016/j.ijhydene.2012.05.129

70. Puthiyapura VK, Pasupathi S, Su H, Liu X, Pollet B, Scott K (2014) Investigation of supported IrO_2 as electrocatalyst for the oxygen evolution reaction in proton exchange membrane water electrolyser. Int J Hydrogen Energy 39:1905–1913. https://doi.org/10.1016/j.ijhydene.2013.11.056

71. Xu J, Liu G, Li J, Wang X (2012) The electrocatalytic properties of an IrO_2/SnO_2 catalyst using SnO_2 as a support and an assisting reagent for the oxygen evolution reaction. Electrochim Acta 59:105–112. https://doi.org/10.1016/j.electacta.2011.10.044

72. Marshall AT, Sunde S, Tsypkin M, Tunold R (2007) Performance of a PEM water electrolysis cell using $Ir_x Ru_y Ta_z O_2$ electrocatalysts for the oxygen evolution electrode. Int J Hydrogen Energy 32:2320–2324. https://doi.org/10.1016/j.ijhydene.2007.02.013

73. Bessarabov D, Millet P (2018) PEM water electrolysis, vol 1. Academic Press

74. Bessarabov D, Millet P (2018) PEM water electrolysis, vol 2. Academic Press

75. Grigoriev SSA, Millet P, Dzhus KA, Middleton H, Saetre TO, Fateev VN (2010) Design and characterization of bi-functional electrocatalytic layers for application in PEM unitized regenerative fuel cells. Int J Hydrogen Energy 35:5070–5076. https://doi.org/10.1016/j.ijhydene.2009.08.081

76. Yim SD, Park GG, Sohn YJ, Lee WY, Yoon YG, Yang TH, Um S, Yu SP, Kim CS (2005) Optimization of PtIr electrocatalyst for PEM URFC. Int J Hydrogen Energy 30:1345–1350. https://doi.org/10.1016/j.ijhydene.2005.04.013

77. Ioroi T, Kitazawa N, Yasuda K, Yamamoto Y, Takenaka H (2001) IrO_2-deposited Pt electrocatalysts for unitized regenerative polymer electrolyte fuel cells. J Appl Electrochem 31:1179–1183. https://doi.org/10.1023/A:1012755809488

78. Chen G, Zhang H, Cheng J, Ma Y, Zhong H (2008) A novel membrane electrode assembly for improving the efficiency of the unitized regenerative fuel cell. Electrochem Commun 10:1373–1376. https://doi.org/10.1016/j.elecom.2008.07.002

79. Tsypkin MA, Lyutikova EK, Fateev VN, Rusanov VD (2000) Catalytic layers in a reversible system comprising an electrolyzing cell and a fuel cell based on solid polymer electrolyte. Russ J Electrochem 36:545–548. https://doi.org/10.1007/BF02757419

80. Zhang Y, Wang C, Wan N, Mao Z (2007) Deposited RuO_2–IrO_2/Pt electrocatalyst for the regenerative fuel cell. Int J Hydrogen Energy 32:400–404. https://doi.org/10.1016/j.ijhydene.2006.06.047

81. Ioroi T, Oku T, Yasuda K, Kumagai N, Miyazaki Y (2003) Influence of PTFE coating on gas diffusion backing for unitized regenerative polymer electrolyte fuel cells. J Power Sources 124:385–389. https://doi.org/10.1016/S0378-7753(03)00795-X

82. Chen G, Zhang H, Ma H, Zhong H (2009) Effect of fabrication methods of bifunctional catalyst layers on unitized regenerative fuel cell performance. Electrochim Acta 54:5454–5462. https://doi.org/10.1016/j.electacta.2009.04.043

83. Altmann S, Kaz T, Friedrich KA (2011) Bifunctional electrodes for unitised regenerative fuel cells. Electrochim Acta 56:4287–4293. https://doi.org/10.1016/j.electacta.2011.01.077

84. Wang YY, Leung DYC, Xuan J, Wang H (2016) A review on unitized regenerative fuel cell technologies, part-A: Unitized regenerative proton exchange membrane fuel cells. Renew Sustain Energy Rev 65:961–977. https://doi.org/10.1016/j.rser.2016.07.046

85. Fiala R, Vaclavu M, Vorokhta M, Khalakhan I, Lavkova J, Potin V, Matolinova I, Matolin V (2015) Proton exchange membrane fuel cell made of magnetron sputtered Pt–CeOx and Pt–Co thin film catalysts. J Power Sources 273:105–109. https://doi.org/10.1016/j.jpowsour.2014.08.093

86. Khalakhan I, Vorokhta M, Kúš P, Dopita M, Václavů M, Fiala R, Tsud N, Skála T, Matolín V (2017) In situ probing of magnetron sputtered Pt-Ni alloy fuel cell catalysts during accelerated durability test using EC-AFM. Electrochim Acta 245:760–769. https://doi.org/10.1016/j.electacta.2017.05.202

Chapter 2
Experimental

This chapter gives a brief insight into the working principles of experimental techniques and methods used during the scientific investigation. A short description of the individual apparatuses is also provided.

Experimental catalytic layers were prepared by means of magnetron sputtering.

The thickness of prepared layers and their roughness was determined by atomic force microscopy. Surface structure was observed by scanning electron microscope.

Valuable information about the chemical composition and distribution of the specimens were obtained by energy-dispersive X-ray spectroscopy.

Chemical states of the elements within the prepared complex structures were studied by X-ray photoelectron spectroscopy; several devices differing in energy of light source and hence probing depth were used.

Electrochemical properties and characteristics of experimental layers were investigated predominantly in-cell by carrying out potentiostatic IV measurements and using electrochemical impedance spectroscopy. In addition, three-electron arrangement was used for cyclic voltammetry and for electrochemical atomic force microscopy measurements.

2.1 Magnetron Sputtering

Sputtering, as a method for thin film preparation, belongs to the family of physical vapor deposition (PVD) techniques. In its simplest form, called diode sputtering, two electrodes are located in low-pressure atmosphere of inert working gas (e.g. Ar). Conductive material which is to be deposited, referred to as target, is mounted on the cathode (negative electrode). Sufficiently high voltage difference is applied between the electrodes, resulting in ionization of working gas and ignition of glow discharge. Created ions are accelerated towards cathode/target and upon hitting its surface sputter out the material which subsequently forms a thin film on substrate located above

© Springer Nature Switzerland AG 2019
P. Kúš, *Thin-Film Catalysts for Proton Exchange Membrane Water Electrolyzers and Unitized Regenerative Fuel Cells*, Springer Theses,
https://doi.org/10.1007/978-3-030-20859-2_2

the target. Diode sputtering, however, suffers from relatively low efficiency of gas ionization and therefore low sputtering yield and consequently slow film growth.

In order to overcome these drawbacks a set of permanent magnets are placed under the cathode/target. The presence of a magnetic field in conjunction with electric field within close proximity of the target creates the so-called $\vec{E} \times \vec{B}$ drift. Trajectories of electrons (with velocity $\vec{v}$ and charge q) caught in this drift are spirally bent by the acting of Lorentz force $\vec{F_L} = q\left(\vec{E} + \vec{v} \times \vec{B}\right)$. Bending and prolonging of the electron's flight path results in significant increase of gas ionization, higher density of plasma and thus higher sputtering yield. The modification of diode sputtering described above is known as magnetron sputtering (Fig. 2.1).

Magnetron sputtering offers several undisputed advantages which makes it suitable not only for R&D but also for industrial applications. The most notable is its ability to reproducibly deposit well-defined thin films of almost any material that is available in form of the target. By introducing reactive gases (N_2 or O_2) into the chamber during the sputtering process, even nitride or oxide thin films can be prepared out of single elemental targets (e.g. Ti to TiO_2). Moreover, magnetron sputtering is not limited solely to the conductive materials; by utilizing RF power supplies, non-conductive ceramic materials or polymers can be deposited as well. By operating several deposition sources simultaneously, alloys with specific composition can also be prepared with relative ease.

With respect to the research in thin-film catalysts for PEM fuel cells and electrolyzers, additional key factors are favoring magnetron sputtering. Firstly, direct control of plasma discharge enables to vary the deposition rate in a wide interval. Therefore, the thickness of resultant films can be controlled from several hundreds of nanometers down to merely few nanometers; a necessity when studying low-loading sputtered catalyst. Secondly, the deposition process itself is carried out at room temperature (omitting the moderate heating up of the substrate due to flux of sputtered material hitting its surface). As such, one can use thermally sensitive materials as substrates without causing any damage to it. In our case the sensitive substrate would be the proton exchange membrane.

Two recently built custom sputtering apparatuses were used for preparation of experimental thin-film catalysts. The smaller device is equipped with three 2-in. balanced magnetron guns (TORUS, K. J. Lesker). The second larger device features three 4-in. balanced magnetrons (TORUS, K. J. Lesker). Additionally, it is equipped with a load-lock system, enabling for faster sample exchange and more precise vacuum control. Both systems use a set of mass flow controllers for composing the working atmosphere (Alicat, flow range from 1 to 100 sccm). Four power supplies (one RF and three DC) are available, which enable simultaneous operation of all sputtering sources on either of individual devices.

More information about magnetron sputtering is provided in [1].

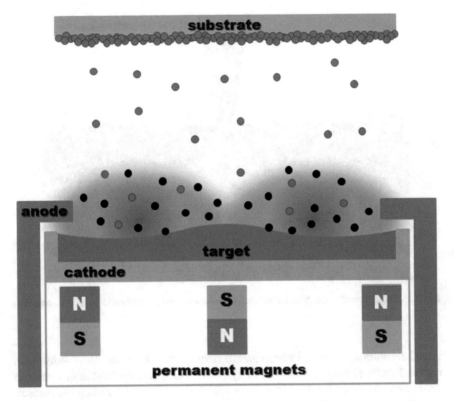

Fig. 2.1 Simplified scheme of magnetron sputtering process

2.2 Scanning Electron Microscopy (SEM)

Scanning electron microscopy (SEM) is a high-vacuum method for morphological investigation of conductive solid state samples. It utilizes a focused high-energy beam of electrons to probe the surface, achieving magnifications up to hundreds of thousands. The primary electrons, emitted by cathode, are accelerated by a series of electromagnetic lenses to several tens of keV. The lenses not only focus the beam but also controllably bend it in order to achieve periodical scanning of the target area, hence the name SEM. Thanks to the short wavelength of the electrons and high degree of their focusation, the beam spot size is in range of few units of nanometers. The primary electrons interact with sample surface in numerous ways, emitting various secondary particles and radiation (see Fig. 2.2) which carry specific information about the impact area.

Secondary electrons (SE) are basically a product of ionization process, induced by primary beam. The kinetic energy of SE is relatively low, less than 50 eV, meaning the inelastic mean free path is in order of nanometers for metals and tens of nanometers for semiconductors. This is the reason why SE mode of observation yields high

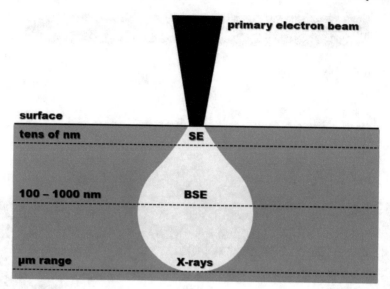

Fig. 2.2 Interaction of primary electron beam with the solid sample and corresponding interaction volume

lateral resolution—detected electrons are only emitted from the close subsurface region directly under the spot of primary beam. The image contrast in this mode is predominantly given by the angle under which the primary beam hits the surface. The larger the angle of incidence, the bigger the volume of which the SE escape to vacuum and thus higher their number. Steep edges therefore appear brighter than planes which are perpendicular to the primary electron beam (see Fig. 2.3).

Backscattered electrons (BSE) are high-energy electrons of primary beam which underwent elastic scattering on the specimen atoms. Since they emerge from much deeper locations of interaction volume, the resolution of BSE images is less than SE images. The BSE imaging however provides valuable information about the elemental composition of the sample. Heavy elements with a higher atomic number backscatter more electrons in comparison to lighter elements. Areas richer in heavy elements therefore appear relatively brighter.

Different contrast interpretations of SE and BSE images have to be kept in mind when observing multi-elemental morphologically complex specimens. The best results are obtained when using both modes simultaneously in order to minimize ambiguity.

Characteristic X-rays are emitted when the primary beam removes the inner shell electron from the specimen atom, causing the vacant shell to be afterwards filled by higher energy electron while releasing excess energy in form of Röntgen radiation. Since each element has a unique set of energy levels, the X-rays generated by the herein described intra-shell transition have specific frequencies. Energy-dispersive X-ray spectroscopy (EDX) is an analytical technique for detection of X-rays and

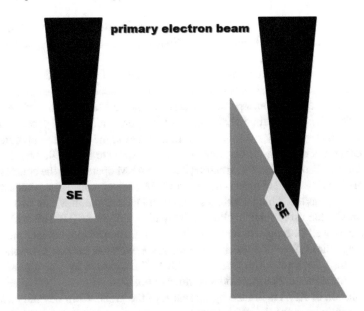

Fig. 2.3 Steep edges appear brighter in SE imaging mode due to larger amount of detected secondary electrons

measuring their energy in order to determine the exact elemental composition of the sample. EDX modules are commonly part of SEM apparatus.

Our setup consists of field emission SEM (Mira III, Tescan), capable of beam energies up to 30 keV and working distances as low as 2 mm. It is able to simultaneously acquire secondary electron as well as backscattered electron images. Also, EDX module (Bruker XFlash® 6|10) is attached, allowing for elemental analysis in form of spectra and maps.

More details about SEM can be found in [2].

2.3 Atomic Force Microscopy (AFM)

Atomic force microscopy (AFM) is an ambient-pressure technique for investigation of various surface properties of solid samples under high magnifications. Besides giving topographical information about the specimen, it is also capable of measuring surface potential distribution, local magnetization, surface conductivity and more. The following lines, however, only describe basic topographic regime of operation.

Simply put, topographic image is obtained by scanning over the sample surface with a sharp tip, attached on a flexible cantilever. Force interaction between the tip and the sample lead to cantilever deflections which are monitored by laser reflection on the quadrant photodiode. Based on the diode readings, the feedback loop controls the piezoelectric elements which adjust the distance between the sample and the tip to keep it constant. Signal from the photodiode is at the same time sent to the PC for raw image reconstruction.

The AFM is capable of operation in several modes, differing in mean tip to surface distance and in nature of interatomic forces being dominant under such conditions. The so-called Lennard-Jones potential describes the course of tip to surface interaction as a function of their mutual distance. When approaching the surface, the tip first feels the attractive van der Waals forces; in this region the AFM can operate in the non-contact mode. Upon further approach, the tip starts to be repelled by the acting of short-range Pauli exclusion principle; here the AFM operates in the contact mode. Arguably the most widely used is the intermittent contact mode (sometimes referred to as tapping mode), which combines the advantages of the previous two regimes. In this mode, the tip oscillates with relatively high amplitude (10–100 nm) through attractive as well as repulsive region of interaction. Intermittent contact mode does not damage the surface of a specimen to the same extent as the contact mode and at the same time does not suffer from image distortions caused by moisture as in case of non-contact mode. For more details, refer to Fig. 2.4.

Resolution of AFM is limited by tip radius of the probe. Commercially available probes usually have the tip radius in order of units of nanometers, lateral resolution is therefore in the same order of magnitude. Vertical resolution, on the other hand, can be achieved even higher; in tenths of nanometers. It should be emphasized that AFM places no demand on conductivity of the samples, as such semiconductors and isolators can be measured in the same way as metals.

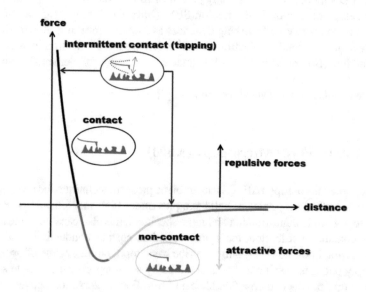

Fig. 2.4 Derivative of Lennard-Jones potential and different AFM operation modes

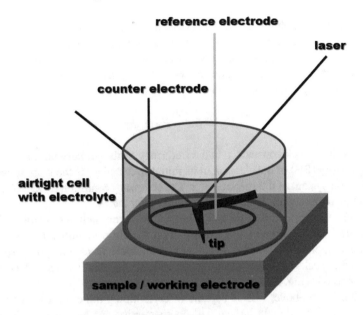

Fig. 2.5 Setup of electrochemical atomic force microscope

Moreover, certain modern variants of AFM are able to combine morphological investigation with electrochemical analysis. These so-called electrochemical AFMs (EC-AFMs) feature glass cap with an O-ring on the bottom which when put on a sample forms a very small airtight in situ electrochemical cell filled with electrolyte. The typical three-electrode configuration is used with reference and counter electrode being inserted to the cell from the top and sample being contacted as working electrode (see Fig. 2.5). This setup allows for investigation of potential-induced topographical changes of the sample directly in the liquid environment of the electrolyte.

Our measurements were carried out on two microscopes. AFM Veeco di MultiMode-V was used for basic topographical analysis, mainly film thickness determination. More complex electrochemical experiments were done using Bruker MultiMode-VIII EC-AFM. Additional information regarding the scanning mode, type of tip, sample setup etc. is given further in the text, when discussing individual experiments.

More information regarding AFM can be found in [3].

2.4 Photoelectron Spectroscopy (PES)

Photoelectron spectroscopy (PES) is an integral, ultra-high vacuum, surface-sensitive method for qualitative and to some extent even quantitative analysis of solid state specimens. It is based on the detection of electrons, emitted from the illuminated part of the sample due to photoelectric effect (usually using soft X-ray). The energy spectrum of these photoelectrons carries valuable information about the sample composition, chemical states of the elements within and their approximate ratio. Considering the law of conservation of energy, the kinetic energy E_K of the elastically emitted photoelectron is given by the equation:

$$E_K = h\upsilon - E_B - \phi_S \qquad (2.1)$$

where h is the Planck constant, υ is the frequency of the primary photon, E_B is the binding energy (BE) of the electron in its initial state and ϕ_S is the work function of the detection system of the spectrometer. The necessary condition for Eq. (2.1) to be valid is the alignment of detector and sample Fermi levels.

The BEs are characteristic for each element's electron shell, the position of peaks in measured photoelectron energy distribution therefore identifies the specimen's elemental composition. Besides detecting true photoelectrons, the typical spectrum also contains peaks of Auger electrons which have E_K that is independent of υ of primary photons (Fig. 2.6).

As mentioned above, PES is capable of determining the chemical states of the elements within the sample surface. The photoemission process generates a vacancy in the atom, which acts as a positive charge and decreases the E_K of photoelectron upon its escape to the vacuum. The extent of this interaction is affected by the electron's density of states, which is in turn influenced by the chemical bonds of the atom with its surroundings. Small shift of the spectral lines, caused by specific photoelectron retardation can be therefore assigned to the certain chemical states of present elements.

The quantitative analysis, i.e. abundances of the elements in flat, homogeneous specimen, is realized by comparing the peak areas divides by the relative sensitivity factors (RSF) which in theory should include all relevant factors, determining the signal's intensity (such as photoionization cross-section, emission angle, transmission function of detector and others).

The surface sensitivity of PES is a result of the fact that emitted photoelectrons have a relatively short inelastic mean free path (IMFP). The IMFP is a function of

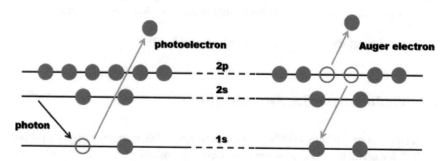

Fig. 2.6 Different mechanisms of true photoelectron and Auger electron emission processes

E_K which is, referring to Eq. (2.1), given by $h\upsilon$. Most laboratory PES apparatuses utilize Mg Kα (1253.6 eV) or Al Kα (1486.6 eV) lines for photoionization; corresponding information depth is then in a range of unites of nanometers up to cca 10 nm (depending on material). This variant of PES is referred to as X-ray photoelectron spectroscopy (XPS).

To obtain even more surface sensitive data it is necessary to lower primary energies. The solution is to use synchrotron radiation photoelectron spectroscopy (SRPES). Not only does it allow the tuning of primary beam energy but it also offers much higher intensities compared to laboratory X-ray sources.

Our experiments were carried out both on laboratory XPS, using a monochromatized Al Kα X-ray source and on SRPES (Elettra-Sincrotrone Trieste) with much lower energies and higher intensities. Both systems feature PHOIBOS 150 hemispherical energy analyzer. Details regarding individual measurements are given further in the text.

For more information about PES, refer to [4].

2.5 Electrochemical Measurements and Characterizations

Various electrochemical methods were employed in order to measure the performance and stability of experimental catalytic layers. The majority of measurements were done in-cell, meaning prepared catalysts were part of fully functioning MEA, sandwiched between bipolar plates at constant pressure, held at constant elevated temperature and supplied by reactants. The cell itself was contacted in four-point probe setup, with separated voltage and current circuits, for more accurate IV readings. The reference electrode was always put to the voltage socket of hydrogen electrode.

Two different types of in-cell measurements were carried out. Firstly, basic DC IV curves were obtained to evaluate the performance and efficiency of the system. Additionally, long-term constant voltage routines yielded information about the stability of MEA. Secondly, more complex potentiostatic electrochemical impedance spectroscopy (PEIS) measurements provided a useful insight into the kinetics of the individual half-reactions and cell resistance.

PEIS is based on applying constant potential with superposed alternating voltage of a certain amplitude (typically around 10 mV) and simultaneous measuring of the current response of the system. The AC frequency changes during the measurement hence the output of the procedure is the dependence of complex impedance on the frequency of the applied AC voltage. The plot of the imaginary part of this complex impedance against its real part is called Nyquist diagram (example of Nyquist plot can be seen in Fig. 2.7).

Interpretation of PEIS spectra is nontrivial, yet certain conclusions can be made relatively simply. Considering Randles model [5], the high frequency intercept on the real axis corresponds to the overall ohmic cell resistance, while the size of the semi arc in the negative imaginary half-plane reflects the kinetic of the reactions.

Fig. 2.7 Example of
Nyquist plots representing
two differently performing
MEAs (black MEA has
faster kinetics)

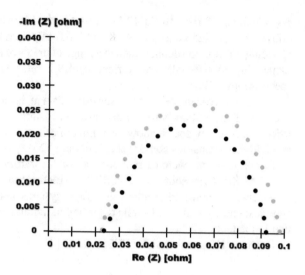

Reactions with faster kinetic (HER, HOR) tend to have much smaller loops than slower reactions (OER, ORR). As such PEIS of PEM-WE and hydrogen PEM-FC plotted in Nyquist diagram usually features just one semi arc, corresponding to the slower reaction; the loop of the faster reaction is simply too small to be recognized.

The in-cell electrochemical measurements described above characterize MEA as a whole. In case the properties of an individual catalyst or structure need to be investigated, one can turn to the three-electrode electrochemical analysis. In this setup the specimen is attached to the working electrode and submerged to the electrolyte, which resembles properties of the actual working environment.[1] Two other electrodes are put to the electrolyte, the reference electrode (in our case Ag/AgCl) and counter electrode (Pt). The connected potentiostat then controls the current flow between counter electrode and working electrode so that the potential difference between working electrode and reference electrode is kept constant. Various experiments can be carried out in this configuration. Potentiostatic experiments can mimic steady operational conditions of certain half-reaction and in conjunction with other methods, such as PES, can be used to evaluate the stability and corrosion resistance of given material. In turn, potentiodynamic experiments, specifically cyclic voltammetry (periodical continuous increase and decrease of applied voltage within fixed interval), simulate repetitive switching on/off of the cell. The obtained voltamograms give insight into the reversible or irreversible nature of redox reactions, taking place on the surface of examined specimen. Additionally, voltamograms can provide relative information about the active surface and activity of catalysts.

All of above-mentioned measurements were controlled by potentiostat Bio-Logic SP-150.

---

[1] H_2SO_4 aqueous solution is usually used to simulate the solid electrolyte of PEM-FC and PEM-WE [6].

References

1. Braun M (2015) Magnetron sputtering technique. In: Handbook of manufacturing engineering technology. Springer, pp 2929–2957. https://doi.org/10.1007/978-1-4471-4670-4_28
2. Amelinckx S, van Dyck D, van Landuyt J (2008) Scanning electron microscopy. In: Handbook of microscopy. Wiley, pp 539–561. https://doi.org/10.1002/9783527620524.ch1
3. Mironov V (2004) SPM textbook. https://sites.google.com/site/vmironovipm/SPM-textbook. Accessed 24 Jan 2018
4. Moulder JF, Chastain J, King RC (1995) Handbook of X-ray photoelectron spectroscopy: a reference book of standard spectra for identification and interpretation of XPS data. Phys Electron
5. Yuan X, Wang H, Colin Sun J, Zhang J (2007) AC impedance technique in PEM fuel cell diagnosis—a review. Int J Hydrogen Energy 32:4365–4380. https://doi.org/10.1016/j.ijhydene.2007.05.036
6. Liu RS, Zhang L, Sun X, Liu H, Zhang J (2011) Electrochemical technologies for energy storage and conversion. Wiley

Chapter 3
Results

The main part of this thesis is divided into several thematic sections. Firstly, the setting up of our PEM-WE testing cell is described (Sect. 3.1) and the way of catalyst loading determination is explained (Sect. 3.2).

After that follows the most extensive section, Sect. 3.3, revolving around magnetron sputtered thin-film low-loading catalysts for anode side of the PEM-WE cell.

Section 3.4 is dedicated to the investigation and optimization of bifunctional thin-film catalysts for PEM-URFC.

Finally, Sect. 3.5 compares the round-trip efficiencies obtained by coupling of dedicated single-purpose MEAs (referential as well as experimental ones) with the efficiencies yielded by experimental bifunctional MEAs.

3.1 PEM-WE Testing Cell Setup

First step in our experimental endeavor was to create an experimental cell for PEM-WE measurements. As was mentioned previously, we have been working in field of PEM-FC for several years and have had graphite single cells (see Fig. 3.1) but not dedicated PEM-WE cells.

Since carbon is undesirable material on the anode side of PEM-WE cell, prone to corrosion under higher potentials, it has to be substituted with more durable alternative. The choice fell on titanium. Question arose whether to keep the same geometry of the flow meanders; after all the reactant on PEM-WE anode is liquid in contrast to gaseous fuel of PEM-FC. Considering the published comparison of different types of flow channels within moderately larger 25 cm^2 cell, [1], we decided to keep the standard serpentine-single channel design. It was reported that this geometry is best for PEM-FC and the performance difference in PEM-WE is negligible to other two channel shapes (see Fig. 3.2). As such serpentine-single channel design is also the optimal variant for the future PEM-URFC applications.

© Springer Nature Switzerland AG 2019
P. Kúš, *Thin-Film Catalysts for Proton Exchange Membrane Water Electrolyzers and Unitized Regenerative Fuel Cells*, Springer Theses,
https://doi.org/10.1007/978-3-030-20859-2_3

Fig. 3.1 Graphite current collectors (end plates) from Greenlight Innovation (active area 2.1 × 2.1 cm²)

(a) **(b)** **(c)**

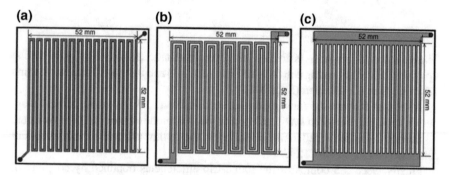

Fig. 3.2 Different flow channels tested for oxygen side of PEM-FC and PEM-WE, **a** serpentine-single channel, **b** serpentine-double channel, **c** parallel channel (reproduced from [1] with permission from ECS)

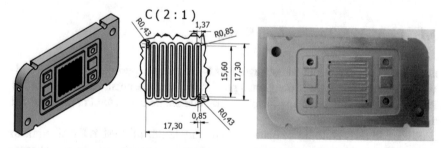

Fig. 3.3 Ti anode end plate with the same dimensions as the original graphite one from the Greenlight Innovation; from left to right: model in Inventor, detail and dimensions of the flow meander (within 2.1 × 2.1 cm² active area), manufactured Ti end plate

Consequently, flow field/current collector (in case of single cell—the end plate) with identical dimensions to those of original Greenlight Innovation end plate was redrawn in Autodesk Inventor and cut from Ti block using a CNC machine (Fig. 3.3).

As was stated in the Sect. (1.3), the corrosion resistance of Ti is due to its ability to rapidly form a layer of TiO_2 on its surface. However, TiO_2 is an insulator. In other words, the Ti end plate has to be additionally coated to prevent formation of the oxide layers in order to prevent the loss of conductivity. Most research groups simply coat it with gold or platinum. The central theme of this work, however, is the maximum decrease in use of noble metals. Using several times more loading for coating of anode end plate than for actual catalysis does not seem right. Therefore, we turned our attention to more cost-efficient alternatives. In collaboration with industry partner Staton we tested two prospective arc-deposited Ti-based coatings, namely TiCrN and TiN. 5 μm thick films were arc deposited on two Ti anode end plates.

Individual anode end plates were then tested in the following way:

1. A Nafion membrane was inserted in between the coated anode end plate and standard graphite cathode end plate
2. Water was supplied to the anode side of the assembled cell as it would be in the case of PEM-WE
3. Constant voltage of 1.7 V (vs. reference electrode on cathode) was applied on cell for 1 h
4. Cell was disassembled and anode end plates was inspected (see Fig. 3.4).

Upon inspection, we concluded that TiCrN layer is not suitable as a corrosion resistant layer. On the other hand, TiN layer showed only moderate signs of corrosion and remained conductive after the experiment. This is in accordance with some of the earlier studies, which evaluated the corrosion protective properties of TiN coating for use in PEM-FC as well as PEM-WE [2, 3]. Based on the underwent experiment and available literature we decided to use TiN as our protective layer for Ti anode end plate, which in pair with standard Greenlight graphite end plate for cathode formed our PEM-WE testing cell.

Rest of the experimental setup included press unit for cell to maintain isobaric conditions as well as PID-controlled heating cartridges to keep the operational temperature constant (see Fig. 3.5). The flow of deionized water[1] to the anode side of

Fig. 3.4 Anode end plates after 1 h at 1.7 V, left—TiCrN layer on Ti; right—TiN layer on Ti

---

[1] 18.2 MΩ cm at 25 °C obtained from Merck Direct Q 3 UV purificator.

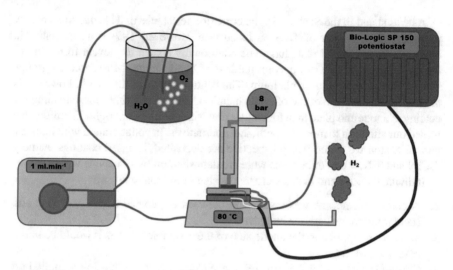

Fig. 3.5 Experimental PEM-WE testing setup

the cell was provided by peristaltic pump. In case, it is not written otherwise, the measuring conditions were as follows[2]:

Press pressure: 8 bar; Temperature: 80 °C; Water flow: 1 ml min^{-1}.

Potentiostat Bio-Logic SP-150 was connected in four-terminal arrangement, where anode voltage probe served as a working electrode and cathode voltage probe as a reference and counter electrode.

3.2 Thin-Film Catalyst Deposition and Noble Metal Loading Determination

All thin-film catalytic layers were prepared by means of magnetron sputtering, using TORUS circular sputtering sources (K. J. Lesker). The target to substrate throw distance was approximately 15 cm. The purity of targets was 99.99%. Prior to the deposition, the chamber was evacuated down to 7.10^{-4} Pa. The sputtering of purely metal thin films was carried out in 0.5 Pa of Ar atmosphere (6.0, Linde) in constant power mode.

In order to precisely determine the loading of thin-film magnetron sputtered noble metal catalysts, we employ the method of nail polish drop.[3] A small drop of nail polish is applied on the surface of silicon wafer and let to dry. The wafer is then placed right

---

[2]The experimental and measuring conditions of PEM-URFC in FC mode are summed up later in text, in Sect. 3.4.1.

[3]Standard gravimetric methods are not ideal due to the very low mass increment after the thin film deposition.

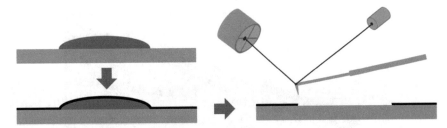

Fig. 3.6 Determination of loading of sputtered material using AFM and the nail polish drop technique

next to the sample which is about to be coated in the magnetron sputtering apparatus. After the deposition, the polish drop (basically in a role of a masking medium) is carefully scratched off which results in the formation of a sharp edge on the interface of Si and sputtered film. The height of this edge can be consequently measured by AFM (see Fig. 3.6). Considering the homogeneous thickness of the layer over the surface of the sample and tabulated density of sputtered material, one can calculate the loading (mass per area). AFM was operating in tapping mode with ScanAsyst Air probes from Bruker which feature 2 nm tip radius.

3.3 Thin-Film Magnetron Sputtered Anode Catalyst for PEM-WE

This subchapter is dedicated to the investigation of magnetron sputtered thin-film low-loading catalyst (less than 0.3 mg cm^{-2} of noble metal) for the anode part of the cell (OER reaction). Our goal was to find out whether sputtering is suitable method for low-loading catalyst deposition and whether we can achieve efficiencies comparable to those presented as state-of-the-art. Speaking of state-of-the-art, Fig. 3.7 compares multiple IV curves reprinted from several recent publications [4], providing a rough reference for performance of a modern PEM-WE. The loading corresponding to plotted curves ranges from 1.2 to 2.5 mg cm^{-2} for Ir on anode and from 0.2 to 0.8 mg cm^{-2} for Pt on cathode.

We decided to set two benchmark points, obtained by averaging the published data, 1 A cm^{-2} at 1.7 V and 2 A cm^{-2} at 1.9 V, to help the eye when comparing our IV curves to performances plotted in Fig. 3.7. Also, the US DOE sets the system energy efficiency target for 2020 to 1 A cm^{-2} at 1.67 V [5], which we can for the sake of clarity conveniently round up to 1.7 V. Achieving aforementioned values with thin-film low-loading anode catalyst was set as our highest challenge.

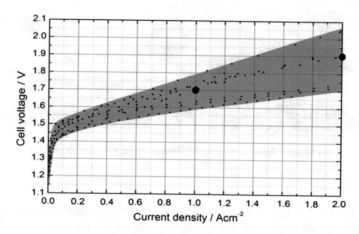

Fig. 3.7 Performance comparison of several recently published polarization curves of PEM-WE single cells, operating at 80 °C (reproduced from [4] with permission from Elsevier); black benchmarks 1 A cm^{-2} at 1.7 V and 2 A cm^{-2} at 1.9 V will serve as a reference points throughout the rest of the thesis

3.3.1 Ir Thin-Film Catalyst Sputtered Directly on Membrane

Our first attempt to the test thin-film anode catalyst involved simple sputtering of 50 nm of pure Ir on a membrane commonly used in PEM-WE. Also, in order to minimize the number of variable parameters, we used commercially available carbon GDL with MPL containing Pt catalyst (hence GDE) for the cathode side of the cell. All relevant parameters of the measured MEA are listed in Table 3.1.

MEA components of the cell before assembly are shown in Fig. 3.8 and details of individual GDLs/current collectors in Fig. 3.9.

After stabilization of the cell temperature, we set the potentiostat to constant voltage of 1.7 V to break-in the catalysts. The observed current density at 1.7 V was very unstable and immediately began to drop drastically. After less than an hour, the current density reached the stable minimum at the value of approximately 80 mA cm^{-2} and did not increase for another 2 h of testing. Such value is below anything which can be considered as efficient. Also, the cell resistance obtained from PEIS spectra (at 1.5 V DC with superposed AC with 5 mV sinus amplitude

Table 3.1 Details of **MEA—Ir sputtered directly on PEM**

Cell	TiN-coated Ti anode end plate, graphite cathode end plate (active area 4.4 cm^2)
Membrane	Nafion N 115 (127 μm thick)
Anode side	50 nm of Ir (113 μg cm^{-2}) sputtered directly on anode side of the membrane Ti mesh GDL (FuelCellsEtc) + 50 nm of Pt sputtered on top of it
Cathode side	Commercial GDE with 0.4 mg cm^{-2} of Pt (Alfa Aesar)

Fig. 3.8 MEA with Ir catalyst sputtered directly on the membrane

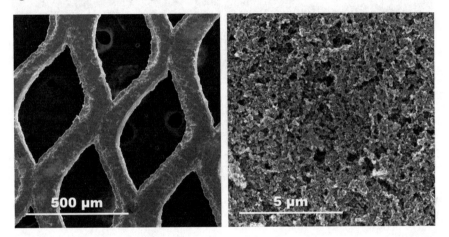

Fig. 3.9 Left—SEM micrograph of Pt-coated Ti mesh GDL; Right—SEM micrograph of commercial GDE with 0.4 mg cm^{-2} of Pt

and frequency from 200 kHz to 500 MHz)[4] was around 0.07 Ω which is above commonly reported values for MEAs with N 115 PEM. Since no actions, such as changing the cell pressure or water flow, lead to an increase of current density, we decided to disassemble the cell to try to find some explanation for the observed poor performance.

The problem was clearly caused by the anode catalyst, i.e. Ir sputtered directly on the membrane. The central part of the membrane, where water is introduced and OER happens, was basically transparent, hinting at poor adhesion of Ir. However, a couple of minutes later, after the membrane dried, the central part started to appear silvery again, proving that, contrary to our first assumption, the catalyst is still

---

[4]All of the PEIS measurements mentioned from this point onward were done keeping the same parameters.

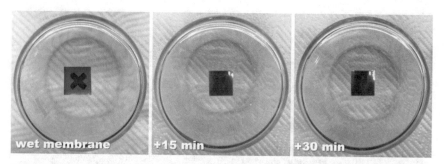

Fig. 3.10 Shrinkage of wet Ir-coated membrane (transparent part becomes silvery again)

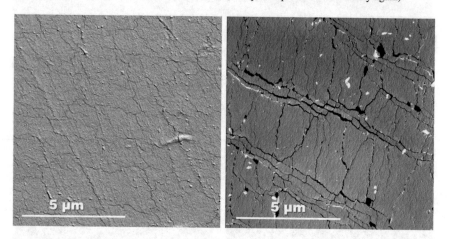

Fig. 3.11 SEM micrographs of as-deposited Ir-coated membrane (left) and membrane after hydration and consequent drying (right)

present (for illustration, see demonstration in Fig. 3.10 and SEM images in Fig. 3.11). Nonetheless, the water uptake of the membrane and its consequent swelling during the PEM-WE operational conditions undoubtedly caused severe rupture of the Ir thin film and loss of its lateral conductivity which is the reason for low OER activity. After all, upon full hydration the N 115 PEM laterally expands by 15% and H_2O represents around 35 wt% of its mass [6]. To sum up, the conducted experiment ruled out the usage of thin-film catalyst sputtered directly on the PEM surface.

3.3.2   Ir Thin-Film Catalyst Sputtered on Ti Mesh GDL

In order to further support our conclusions from the 1st experiment, we now deposited the Ir catalyst on top of the Ti mesh instead of PEM. Ti mesh is rigid and does not stretch when exposed to water, hence the conductivity and integrity of the thin film should not be disrupted and the performance of the cell should be stable. MEA details

Table 3.2 Details of **MEA—Ir sputtered on Ti mesh GDL**

Cell	TiN-coated Ti anode end plate, graphite cathode end plate (active area 4.4 cm^2)
Membrane	Nafion N 115 (127 μm thick)
Anode side	50 nm of Ir (113 μg cm^{-2}) sputtered on Ti mesh GDL (FuelCellsEtc)
Cathode side	Commercial GDE with 0.4 mg cm^{-2} of Pt (Alfa Aesar)

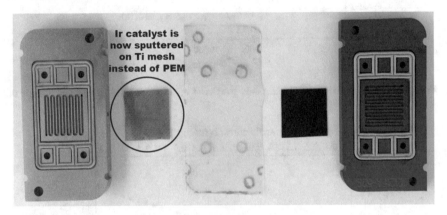

Fig. 3.12 MEA with Ir catalyst sputtered on top of Ti mesh GDL

are in Table 3.2. The changes of cell MEA setup in respect to the previous experiment are emphasized in Fig. 3.12.

Indeed, after assembling the cell and starting the constant voltage 8.5 h break-in procedure at 1.7 V,[5] we witnessed fairly stable values of current density of around 270–280 mA cm^{-2} (see Fig. 3.13). The consequently measured IV curve obtained in stepped galvanostatic mode[6] can be seen in Fig. 3.14.

Analyzing the data, it could be concluded that performance of the cell is poor, the current densities at 1.7 and at 1.9 V are way below our reference points (benchmarks do not even fit into the scale of the graph). Nonetheless, the performance was stable and reproducible (similar IV curves were obtained the next day and the day after), proving that the thin-film Ir catalyst was not washed off the mesh or rendered inactive in any way. Cell resistance, determined from PEIS curves, went down in comparison to the previous experiment from 0.07 to 0.05 Ω. Poor performance of the cell was most likely due to the low area of Ti mesh, the surface of which Ir was sputtered on. For current densities to go up, we needed to find suitable high-surface catalyst support.

---

[5]When referring to break-in procedure later in the text, we will always mean constant voltage of 1.7 V for duration of 8.5 h.

[6]Applying 100 mA steps with 15 s stabilization period. From Sect. 3.3.4 onwards the stepped galvanostatic mode was due to the instrumental reasons (random occurrences of sharp high voltage spikes) replaced by finer stepped potentiostatic mode (applying 5 mV steps with 10 s stabilization period).

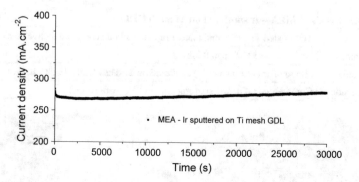

Fig. 3.13 Break-in procedure of **MEA—Ir sputtered on Ti mesh GDL** at constant voltage of 1.7 V

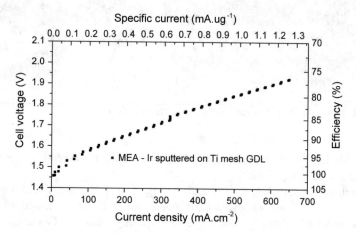

Fig. 3.14 IV curve of **MEA—Ir sputtered on Ti mesh GDL** [Note that upper X-axis represents the specific current, i.e. the current per mass of noble metal catalyst (anode and cathode combined). It proves to be a valuable variable when comparing low-loading catalysts and we will keep using it in all our IV graphs from this point onward. Right Y-axis represents the thermodynamic efficiency calculated using relation (1.14)]

3.3.3 Ir Thin-Film Catalyst Sputtered on Ti-Coated Carbon Paper GDL

It proved to be relatively difficult to find commercially available Ti mesh with pattern finer than the one we originally used. We also tried to experiment with several types of Ti felts but they were generally too rigid and thick and could not be properly placed into our experimental cell. In the end, we decided to improvise and to use high-surface carbon paper GDL, namely Sigracet 29AA, as our next catalyst support. Obviously, carbon paper on its own would not withstand the harsh conditions of the anode side of PEM-WE. However, we believed that sputtering 50 nm of Ti on both

Table 3.3 Details of **MEA—Ir sputtered on Ti-coated carbon GDL**

Cell	TiN-coated Ti anode end plate, graphite cathode end plate (active area 4.4 cm^2)
Membrane	Nafion N 115 (127 μm thick)
Anode side	50 nm of Ir (113 μg cm^{-2}) sputtered on 50 nm Ti-coated carbon paper GDL (Sigracet 29AA)
Cathode side	Commercial GDE with 0.4 mg cm^{-2} of Pt (Alfa Aesar)

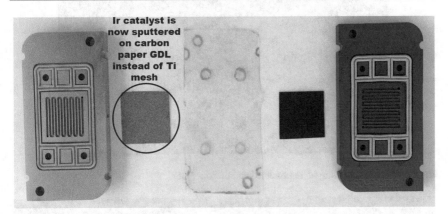

Fig. 3.15 MEA with Ir catalyst sputtered on top of Ti-coated carbon paper GDL

sides of it and consequently 50 nm of Ir on the side closer to the PEM might provide enough corrosion protection to last at least for the duration of experiment. MEA details are summed up in Table 3.3 and changes done to the MEA setup are specified in Fig. 3.15.

Indeed, Ti sputter-coated carbon paper GDLs have been reported elsewhere [7, 8] (albeit not in conjunction with thin-film low-loading Ir catalysts) and have been proven to withstand anode conditions for necessary amount of time (several tens of hours).

The morphological alteration of carbon paper's surface after individual sputtering steps can be seen in Fig. 3.16. Reportedly, Ti sublayer not only prevents the corrosion of carbon but also improves the adherence of catalytic Ir layer.

EDX mapping of coated GDL (see Fig. 3.17) proves homogeneous distribution of both elements. The blurrier nature of green image is presumably due to the double-sided deposition of Ti and the large information depth of the characteristic X-rays, which allow us to partially see the "bottom side" of the GDL.

The IV curve, obtained after break-in procedure, is plotted in grey in Fig. 3.18. The curve is compared to the performance of MEA with Ir catalyst deposited on Ti mesh from previous experiment (black curve). Figure 3.18 also features benchmarks 1 A cm^{-2} at 1.7 V and 2 A cm^{-2} at 1.9 V previously introduced in Sect. 3.3. As expected, using high-surface catalyst support significantly increased cell's current

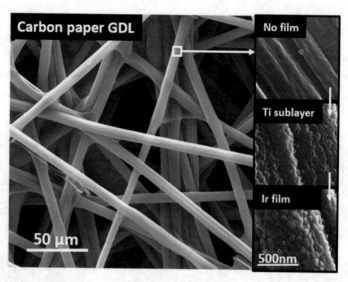

Fig. 3.16 Catalyst coating of Ti-protected carbon paper GDL

Fig. 3.17 EDX mapping of catalyst-coated carbon paper GDL

densities. Also, the cell resistance derived from PEIS measurements was around 0.037 Ω which is better than in case of MEA with Ir-coated Ti meshes.

The overall performance of the cell is now approaching the values more typical for high-loading MEAs. Note however that total catalyst loading (anode + cathode) of our experimental MEA is just 513 μg cm^{-2} which is at least 2–3 times less than the loadings presented in Fig. 3.7. In other words, although the low-loading sputtered catalyst tends to have lower absolute current densities, the specific current is much higher than for high-loading counterparts. The performance was stable and we obtained very similar IV curve after another endurance test of 8.5 h constant voltage at 1.7 V. Even though we did not notice any performance drop over time of our testing, it could be expected that carbon paper would eventually corrode. After all, the Ti layer was deposited by sputtering, which is in principle directional, meaning

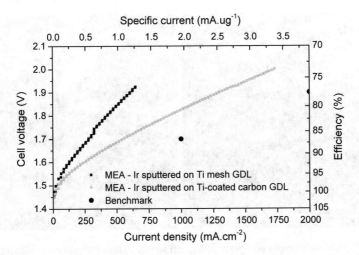

Fig. 3.18 IV curve of **MEA—Ir sputtered on Ti-coated carbon GDL** compared to **MEA—Ir sputtered on Ti mesh GDL**

that there are undoubtedly some places which were shielded from the particle flux and remained uncoated.

Interestingly enough, when we used Ti-coated Sigracet 29BC (see Fig. 3.19), i.e. the carbon paper with additional MPL of carbon black on top of it, believing it would further enhance the performance due to even larger surface than 29AA, the water in the closed circuit of our PEM-WE testing setup started to turn black immediately after setting the potential to 1.7 V and current density went down to zero. Apparently, carbon black microporous layer has so many hollows and creases that it simply cannot be effectively protected by Ti thin film.

To sum-up, this experiment proved that carbon paper GDL without the carbon black MPL can indeed serve as an improvised high-surface short-to-mid-term substitution for Ti-based GDLs, provided it is thoroughly coated with Ti from both sides. Obviously, stability of MEA is one of the crucial factors when talking about commercial devices, but for R&D purposes even short-time comparisons of various experimental catalysts might yield interesting results and that is where Ti-coated carbon paper could be useful.

3.3.4 Ir Supported on TiC Nanoparticles

We have so far demonstrated that direct sputter-coating of PEM membrane does not work well and that coating the GDL is a much better solution. We have seen that the more surface available for the catalyst on the support medium the higher the current densities. In PEM-FC industry, as stated in Sect. 1.2, Pt catalyst is usually supported

Fig. 3.19 SEM micrographs of Sigracet 29BC carbon paper GDL with carbon black MPL before (left) and after (right) the coating with 50 nm of Ti and 50 nm of Ir

on high-surface carbon black nanoparticles mixed with ionomer which are in turn supported on carbon paper (forming GDE) or printed on membrane (forming CCM). The question arises whether similar approach can be mimicked on the PEM-WE anode. Suitable support material has to be found, which will not corrode and will remain conductive. Reflecting on recently published promising results, we chose to work with conductive ceramic, namely with TiC nanopowder [9–11].

The plan was to create a mixture of ceramic nanoparticles and ionomer which would serve as a support material for anode catalyst.[7] Our approach revolves around thin-film deposition; therefore, the catalyst was not to be dispersed within the volume of support material, which is considered to be a more conventional way, but merely on the very top of it, in unprecedentedly low loadings. After all, it is widely accepted that when the catalyst is conventionally dispersed within the volume, only about 40% of it is truly active [12, 13].

Since Ti mesh GDL did not provide enough surface to effectively carry the supported catalyst, the catalyst-coated support material was hot-pressed directly on the membrane (see Fig. 3.20).

Preparation details of TiC-based support sublayer coated with Ir:

A mixture of TiC nanoparticles (99+ %, 40–60 nm, cubic structure, US Research Nanomaterials), Nafion solution (5 wt%, Dupont; 15 wt% in respect to TiC), isopropanol and deionized water (1 and 0.1 ml g^{-1} in respect to TiC) was ultrasonically stirred for 1 h to create the ink of support material. After homogenization, the support ink was evenly spread over the PTFE transient foil (via roller) and left to dry to get rid of the volatile solution for 4 h (resultant loading was 0.4 mg cm^{-2}). PTFE

---

[7]Certain formulations and figures within Sect. 3.3.4 are taken from already published paper written by Kúš et al. [14].

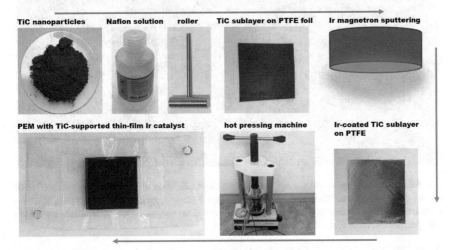

Fig. 3.20 Left—conventional way of supporting catalyst within the whole volume of support material (usual noble metal loading is in mg range); right—our thin-film approach, where catalyst is coated just on the surface of support material (noble metal loading in 0.1 mg range)

Fig. 3.21 Schematic representation of CCM preparation with TiC-supported thin-film Ir anode catalyst ("bottom" geometry)

foil with the sublayer was subsequently put into the magnetron sputtering chamber for deposition of the thin-film Ir anode catalyst (50 nm). Finally, the catalyst-coated sublayer was hot-pressed (120 °C, 150 kg cm^{-2}, 150 s) from transient PTFE foil onto the Nafion N 115 PEM, forming the CCM. The process is schematically described in Fig. 3.21.

The resulting CCM composition is identical to that on the image on the right in Fig. 3.20; since the Ir catalyst is after the hot pressing so-to-say below the TiC-based sublayer, we nicknamed this geometry "bottom".

EDX mapping of the Ir-coated TiC sublayer, done after the Ir deposition and before the hot pressing, proved homogeneous dispersion of ionomer and TiC nanoparticles as well as even coverage by Ir thin film (Fig. 3.22).[8]

After the morphological analysis we proceeded to actual in-cell performance test. Tested MEA can be seen in Fig. 3.23 and its details are listed in Table 3.4.

---

[8]Note that even if not mentioned in text, EDX analysis was routinely carried out on all future samples to confirm homogeneous distribution of elements within.

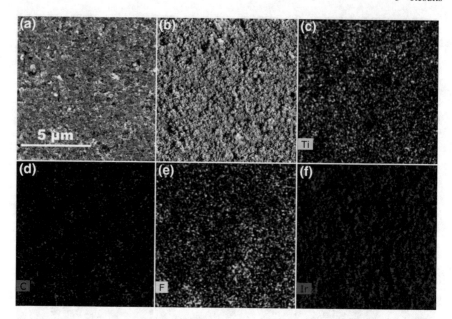

Fig. 3.22 **a** SEM image of 0.4 mg cm^{-2} support sublayer; **b** 50 nm of Ir catalyst sputtered on top of 0.4 mg cm^{-2} support sublayer; **c, d, e, f** EDX elemental mapping of containing elements, obtained from image (**b**). All micrographs were taken with the same magnification (reproduces from [14] with permission from Elsevier)

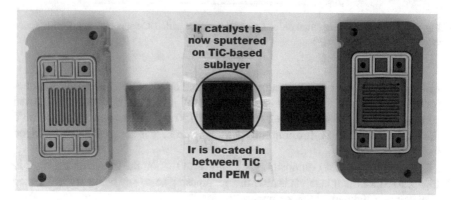

Fig. 3.23 MEA with Ir catalyst supported on TiC ("bottom" geometry)

Table 3.4 Details of **MEA—Ir supported on TiC "bottom"**

Cell	TiN-coated Ti anode end plate, graphite cathode end plate (active area 4.4 cm^2)
Membrane	Nafion N 115 (127 μm thick)
Anode side	50 nm of Ir (113 μg cm^{-2}) sputtered on TiC-based sublayer 0.4 mg cm^{-2} (15 wt% of ionomer), hot-pressed on anode side of the membrane Ti mesh GDL (FuelCellsEtc) + 50 nm of Pt sputtered on top of it
Cathode side	Commercial GDE with 0.4 mg cm^{-2} of Pt (Alfa Aesar)

Fig. 3.24   Left—thin-film MEA in configuration "bottom"; right—thin-film MEA in configuration "top"

Upon starting the standard break-in procedure, we observed very similar behavior as in our first experiment, where we deposited Ir directly onto the PEM (**MEA—Ir sputtered directly on PEM**). Again, the current density went immediately rapidly down, stabilizing itself at negligible values of approximately 40 mA cm^{-2}. However, in this case, the lateral expansion of PEM should not be the main cause of the problem since the catalyst is no longer deposited on membrane but supported on the TiC-based layer which even after expansion should provide sufficient electron conductivity (at least at the beginning of the experiment, before potential corrosion). We speculated that poor performance might be due to anode catalyst reagent starvation—several micrometers thick TiC layer might not be porous enough to provide necessary water flow down to the so-to-say localized TPB by the membrane. Similarly, produced oxygen bubbles might end up being encapsulated under the support layer thus blocking the active sites of the catalyst. Either way, to circumvent the potential mass transport problem, the most logical step seemed to be to modify the preparation method such that Ir thin film would be on the other side of TiC-based support sublayer (i.e. in position "top", see Fig. 3.24).

To achieve this, we simply changed the order of actions within the preparation process, namely we interchanged the Ir deposition and hot pressing (see Fig. 3.25).

Corresponding MEA details are same as in Table 3.4, only the geometry is now "top" as can be seen by the silvery gloss of Ir on the TiC-based sublayer in Fig. 3.26.

Positioning the Ir thin-film catalyst on the outer side of the MEA had significant effect on the performance. The break-in procedure led to current stabilization at about 700 mA cm^{-2}. The IV curve of **MEA—Ir supported on TiC "top"** compared to **MEA—Ir sputtered on Ti-coated carbon GDL** (i.e. the previous best performance) can be seen in Fig. 3.27.

We can see that the efficiency of **MEA—Ir supported on TiC "top"** was higher in the lower potential regions, however in higher potential regions **MEA—Ir on Ti-coated carbon GDL** performed better. The medium to high current density region is mainly affected by the resistance of the flow of ions to the other side of PEM and mass transfer effects [15]. Since the PEM was same in both cases, the limitations of **MEA—Ir supported on TiC "top"** are presumably related to the TiC sublayer; its composition, thickness etc.

Indeed, after comparing the PEIS diagrams (Fig. 3.28), we concluded that although the high surface of TiC nanoparticles definitely boosts the overall kinetics of OER (the semi arc of **MEA—Ir supported on TiC "top"** is much smaller), the actual ohmic resistance of MEA (the high frequency intersection on the real axis)

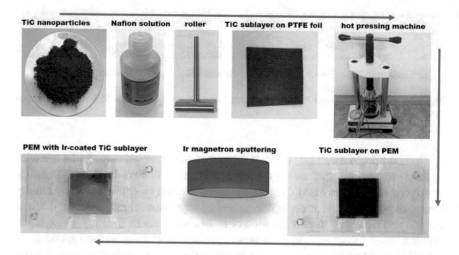

Fig. 3.25 Schematic representation of CCM preparation with TiC-supported thin-film Ir anode catalyst ("top" geometry)

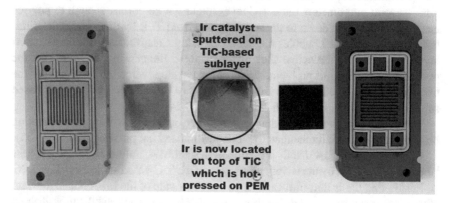

Fig. 3.26 MEA with Ir catalyst supported on TiC ("top" geometry)

is slightly smaller in case of **MEA—Ir on Ti-coated carbon GDL**; 0.037 versus 0.04 Ω to be precise.

Based on the fact that the TPB, necessary for electrochemical water decomposition, is localized solely in the surface region of the support TiC-based sublayer we believe that the subsurface mass of the support material (see illustration in Fig. 3.29) represents merely redundant ionic resistance for protons permeating to the cathode side of the PEM.

To prove our hypothesis, we decided to optimize the TiC-based support sublayer by tuning its overall loading as well as weight ratio of Nafion ionomer within.

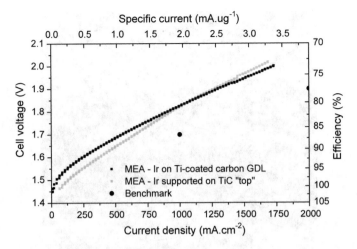

Fig. 3.27 IV curve of **MEA—Ir supported on TiC "top"** compared to **MEA—Ir sputtered on Ti-coated carbon GDL**

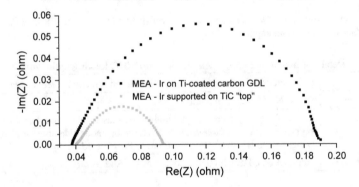

Fig. 3.28 Nyquist diagrams of **MEA—Ir supported on TiC "top"** and **MEA—Ir on Ti-coated carbon GDL** obtained at 1.5 V

Influence of the support material loading on the PEM-WE performance

Firstly, we assembled a series of three MEAs with different total loading of TiC-based sublayer (0.4 down to 0.1 mg cm^{-2}) but the same chemical composition. The MEA details for individual samples are given in Table 3.5.

From the morphological point of view, the surface structure of the support sublayer remains unaltered for all three tested loadings. It is to be expected since varying the loading of support material only affects the bulk thickness of support sublayer not its roughness. As such, SEM micrographs were indistinguishable to that in Fig. 3.22a (before Ir deposition) and Fig. 3.22b (after Ir deposition). In terms of thickness, based on the cross-section imaging (Fig. 3.29), for each 0.1 mg cm^{-2} of TiC-based support material loading the sublayer is approximately 0.5 μm thicker.

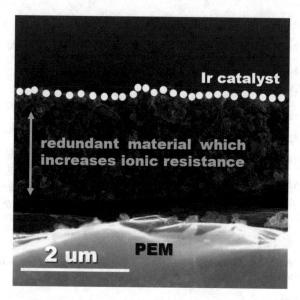

Fig. 3.29 cross-section of TiC-based support sublayer (loading 0.4 mg cm^{-2}); note that just the very surface of the sublayer supports Ir catalyst

Table 3.5 Details of **MEAs—Ir supported on TiC "top", TiC-loading optimization**

Cell	TiN-coated Ti anode end plate, graphite cathode end plate (active area 4.4 cm^2)
Membrane	Nafion N 115 (127 μm thick)
Anode side	50 nm of Ir (113 μg cm^{-2}) sputtered on TiC-based sublayer **0.4** or **0.2** or **0.1 mg cm^{-2}** (15 wt% of ionomer), hot-pressed on anode side of the membrane Ti mesh GDL (FuelCellsEtc) + 50 nm of Pt sputtered on top of it
Cathode side	Commercial GDE with 0.4 mg cm^{-2} of Pt (Alfa Aesar)

The IV curves obtained after break-in procedure can be seen in Fig. 3.30.

Polarization curves, measured for individual MEAs confirm that decreasing the loading of support material indeed leads to better PEM-WE performance. As seen from Nyquist plots (Fig. 3.31), the kinetics of the electrochemical reactions, represented by the semi arcs, do not vary significantly but the overall ohmic cell resistance (real axis intersection) tends to increase with increasing loading (thickness) of support sublayer (0.04 Ω for 0.4 mg cm^{-2}, 0.034 Ω for 0.2 mg cm^{-2} and 0.03 Ω for 0.1 mg cm^{-2}).

This implies that the 50 nm Ir catalyst is utilized approximately equally on all three samples, since the roughness and the surface area of the support remains unaltered, yet the overall PEM-WE performance seems to be hindered by the voltage loss proportional to the thickness (loading) of the support material, forming the sublayer.

Obtained data lead us to believe that in order to maximize the PEM-WE performance, loading (thickness) of the support material, forming the sublayer should be

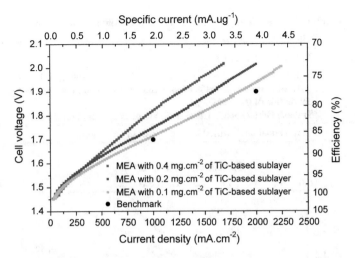

Fig. 3.30 IV curves of **MEAs—Ir supported on TiC "top", TiC-loading optimization**

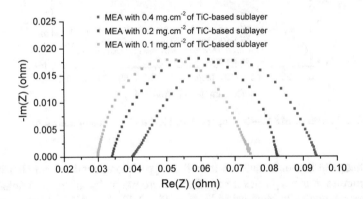

Fig. 3.31 Nyquist diagrams of **MEAs—Ir supported on TiC "top", TiC-loading optimization** obtained at 1.5 V

minimized. Therefore, the loading of 0.1 mg cm^{-2} was regarded as optimal and fixed for the following experiments.[9]

Influence of the ionomer content within the support sublayer on the PEM-WE performance

The next step in the TiC-based support sublayer optimization was to determine the optimal amount of Nafion within. For this experiment we prepared three MEAs with different ratios of ionomer to TiC (30 down to 5 wt%) (see Table 3.6).

---

[9]Lower loadings than 0.1 mg cm^{-2} resulted in undesired discontinuous layers and were not further investigated.

Table 3.6 details of **MEAs—Ir supported on TiC "top", Nafion optimization**

Cell	TiN-coated Ti anode end plate, graphite cathode end plate (active area 4.4 cm²)
Membrane	Nafion N 115 (127 μm thick)
Anode side	50 nm of Ir (113 μg cm⁻²) sputtered on TiC-based sublayer 0.1 mg cm⁻² (**30 wt% of ionomer, 15 wt% of ionomer** or **5 wt% of ionomer**), hot-pressed on anode side of the membrane Ti mesh GDL (FuelCellsEtc) + 50 nm of Pt sputtered on top of it
Cathode side	Commercial GDE with 0.4 mg cm⁻² of Pt (Alfa Aesar)

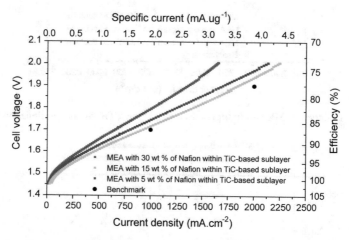

Fig. 3.32 IV curves of **MEAs—Ir supported on TiC "top", Nafion optimization**

A change in ionomer content within the support compound did not alter its surface structure in any significant way. SEM micrographs for all three samples were again morphologically identical to those in Fig. 3.22. Also, EDX mapping proved homogeneous dispersion of the ionomer for all investigated concentrations.

Polarization curves of investigated MEAs, after break-in procedure are plotted in Fig. 3.32.

We noticed considerably worse performance for MEAs containing 30 and 5 wt% of ionomer in comparison to 15 wt%. Interestingly enough, the Nyquist diagram (Fig. 3.33) reveals that unlike in the preceding experiment, where the variation of the support material loading led to overall cell's ohmic resistance changes (Fig. 3.31), in case of the ionomer content alteration, the kinetics itself of the electrochemical reaction (i.e. the size of plotted semi arc) is affected. The MEA with 15 wt% ionomer content exhibits the fastest kinetics (smallest semi arc).

Evidently, certain ideal ratio of TiC conductive nanoparticles and Nafion ionomer within the support sublayer has to be established. Should the ionomer content be too low, ion-conducting paths would be sparsely scattered, resulting in inefficient formation of electrochemically active TPB and slower reaction kinetics. Analogously,

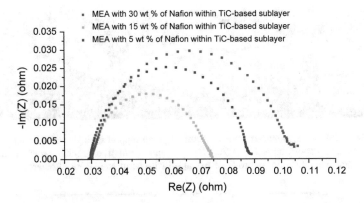

Fig. 3.33 Nyquist diagrams of **MEAs—Ir supported on TiC "top", Nafion optimization** obtained at 1.5 V

the excess of the ionomer results in insufficient electron conductivity of the support material. Similar observations were reported elsewhere [8].

Last but not least, it should be emphasized that Nafion serves not only as ion conductor but also as a binder for TiC nanoparticles. We noticed that success rate of faultless hot press of the 5 wt% ionomer support material from PTFE to N 115 PEM is significantly lower than that of 15 or 30 wt%. Considering this and the fact that the MEA with 15 wt% of ionomer performed best among the three investigated samples, we chose to fix this concentration for further experiments.

Influence of the Ir catalyst loading sputtered on top of the support sublayer on the PEM-WE performance

Based on the results of the previous two experiments, the ionomer content within the support material was optimized to 15 wt% (in respect to TiC) and the support material loading hot-pressed to PEM to 0.1 mg cm^{-2}.

In order to investigate the effect of the amount of sputtered Ir catalyst, located on top of the TiC-based support sublayer, on the PEM-WE performance, three MEAs with different thickness of deposited Ir (see Table 3.7) were tested and their performances compared.

Table 3.7 Details of **MEAs—Ir supported on TiC "top", Ir loading optimization**

Cell	TiN-coated Ti anode end plate, graphite cathode end plate (active area 4.4 cm^2)
Membrane	Nafion N 115 (127 μm thick)
Anode side	**25 nm of Ir (56.5 μg cm^{-2}), 50 nm of Ir (113 μg cm^{-2}) or 75 nm (169.5 μg cm^{-2}) or Ir** sputtered on TiC-based sublayer 0.1 mg cm^{-2} (15 wt% of ionomer), hot-pressed on anode side of the membrane Ti mesh GDL (FuelCellsEtc) + 50 nm of Pt sputtered on top of it
Cathode side	Commercial GDE with 0.4 mg cm^{-2} of Pt (Alfa Aesar)

Fig. 3.34 SEM micrographs (secondary electrons) of the support sublayer before catalyst deposition (**a**) and after deposition of 25 nm (**b**), 50 nm (**c**) and 75 nm (**d**) of Ir (all micrographs are taken with the same magnification) (reproduces from [14] with permission from Elsevier)

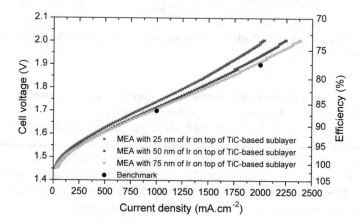

Fig. 3.35 IV curves of **MEAs—Ir supported on TiC "top", Ir loading optimization** (note that upper X-axis is missing because the catalyst loadings are now different for the three tested MEAs)

SEM micrographs show surface morphology alterations of the samples due to the increasing thickness of sputtered Ir catalytic film (Fig. 3.34).

IV curves, obtained after break-in procedure, (Fig. 3.35) prove that increasing the amount of sputtered Ir improves the PEM-WE performance.

The effect is more apparent when the Ir film thickness is increasing from 25 to 50 nm. Increasing thickness from 50 to 75 nm does further improve the performance but less noticeably, especially in low potential region.

PEIS measurements (Fig. 3.36) confirmed that reaction kinetics are indeed faster with higher catalyst loadings (smaller semi arcs for higher Ir loadings).

To properly interpret this outcome, it has to be clarified that prolonging the Ir deposition time results not only in thicker layer of Ir but also in formation of columnar structures which presumably enhance the surface utilization of the catalyst (see Fig. 3.37). Also, not only the area in sputtering source's direct "line of sight" gets coated. Processes such as reflection and scattering of incident particle flux on substrate and resputtering of the deposited material [16] lead to coating of obscured parts of the support in close subsurface region and ultimately to larger TPB and

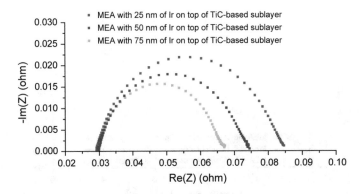

Fig. 3.36 Nyquist diagrams of **MEAs—Ir supported on TiC "top", Ir loading optimization** obtained at 1.5 V

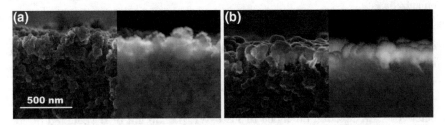

Fig. 3.37 cross-section of the support sublayer with 25 nm (**a**) and 75 nm (**b**) of sputtered Ir on top of it; left part—secondary electrons imaging; right part—backscattered electrons imaging (all micrographs are taken with the same magnification) (reproduces from [14] with permission from Elsevier)

better PEM-WE performance. Both mentioned effects can be seen when comparing cross sections of samples with 25 nm (Fig. 3.37a) and 75 nm (Fig. 3.37b) of the Ir catalyst.

When comparing MEAs with different thickness of Ir thin film on TiC-based sublayer in terms of overall noble metal loading (i.e. anode and cathode catalyst combined), the situation is different (see Fig. 3.38). Now the MEA with 25 nm exhibits the highest specific performance closely followed by MEA with 50 nm and finally the lowest efficiency, predominantly in mid to high current region, shows MEA with 75 nm.

To sum up, based on the curves plotted in Figs. 3.35, 3.38, we conclude that increasing the film thickness of sputtered Ir catalyst supported on TiC-based sublayer on anode from 25 to 50 nm leads to significant increase in absolute PEM-WE performance, yet the specific performance remains nearly the same. Further increasing of the film thickness to 75 nm pushes the absolute efficiency even higher but at the same time the noble metal utilization (current per loading of noble metal) tends to go down. Taking aforementioned into account, we consider 50 nm of Ir ($113 \, \mu g \, cm^{-2}$) to be optimal from the point of view of absolute as well as specific performance.

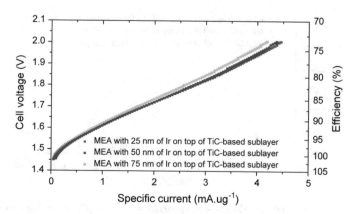

Fig. 3.38 IV curves with specific current of **MEAs—Ir supported on TiC "top", Ir loading optimization**

Chemical analysis of Ir catalyst supported on TiC nanoparticles and stability testing

The previous three comparative experiments helped us to determine optimal parameters of experimental low-loading Ir thin-film catalyst supported on TiC nanoparticles for OER reaction on anode of PEM-WE [i.e. 50 nm of Ir sputtered on top of 0.1 mg cm^{-2} TiC-based sublayer (15 wt% of Nafion in respect to TiC)]. Since the results seemed to be very promising and obtained efficiencies were among the highest we measured so far (actually not too different from the benchmarked efficiencies), we turned our attention to the investigation of chemical state and stability of elements within the anode side of the cell.

We sputtered 50 nm of Ir on Ti foil and carried out an accelerated aging procedure, consisting of three-electrode potential cycling in 0.1 M H_2SO_4 aqueous solution. We used Pt wire counter electrode and leak-free Ag/AgCl (3.4 M KCl) reference electrode. Potential cycling was performed in range from 0 to 1.7 V[10] versus RHE (reversible hydrogen electrode) at 50 mV s^{-1} scan rate for 500 times (last scan was terminated at 1.7 V). After the cycling, the sample was transferred via N_2 inflated glovebox to the XPS apparatus and probed by conventional Al Kα X-ray source (1486.6 eV).

Comparison of the Ir 4f XPS spectra acquired before and after the cycling, with subtracted Shirley-type background and after charge correction to C 1s line at 284.5 eV, can be seen in Fig. 3.39.

Straight after the deposition, the Ir is not 100% metallic (i.e. in state Ir^0) but contains a minor fraction of oxidized species at higher binding energies (Ir^0 and Ir^{4+} were fitted by two pairs of Voight peaks). These are probably present due to the insufficient presputtering evacuation conditions as some residual partial pressure of oxygen might have remained in the chamber. After the cycling, Ir gets fully oxidized

---

[10]This is relevant for simulation of switching on and off of the PEM-WE cell.

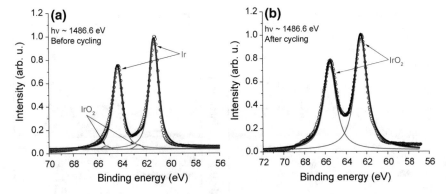

Fig. 3.39 XPS spectra of Ir 4f before (**a**) and after (**b**) the electrochemical cycling (spectra are normalized to the maximum peak intensity)

with no traces of metallic Ir (see Fig. 3.39b). This is proven not only by the fitting of Ir 4f spectra but also by high increment in intensity of O 1s peak.[11] According to the literature, the chemical state IrO_2 is responsible for high activity towards OER. This is one of the reasons why after the assembly of the cell, the MEA has to undergo a certain break-in procedure in order to "activate" the catalysts.

In addition to analyzing anode catalyst, we also turned our attention to the catalyst support. After all, the high surface of TiC-based support sublayer is the main reason why we were able to obtain such high performances with just a fraction of conventional loading of noble metals. As such, it is essential to have information about the chemical stability of TiC under the harsh conditions of PEM-WE anode.

Again, we used Ti foil as a substrate onto which we sprayed 0.1 mg cm^{-2} of the support material (15 wt% ionomer in respect to TiC). Similarly, we performed 500 potential cycles in a range from 0 to 1.7 V versus RHE at 50 mV s^{-1} in 0.1 M H$_2$SO$_4$ aqueous solution (last scan was terminated at 1.7 V). Thanks to access to Material Science Beamline at Elettra-Sincrotrone in Trieste, we were able to measure not only standard XPS spectra but also more surface sensitive SRPES spectra. The near-surface area of the samples was probed, using synchrotron radiation with photon energy of 630.0 eV.

A conventional Al Ka X-ray source (1486.6 eV) was used for gaining the information from deeper layers of the system (XPS).[12] Both SRPES and XPS spectra acquired before and after the cycling (transfer was again done through N$_2$ inflated glovebox) can be found in Fig. 3.40.

The complex Ti 2p spectra were fitted by four pairs of Voight peaks [18, 19] after subtraction of a Shirley-type background and charge shift correction to C 1s line at 284.5 eV.

---

[11] It is fair to expect that O 1s signal is from iridium oxide and not from the Ti foil since no Ti 2p peaks are visible in the XPS spectra (both before and after the aging procedure).

[12] According the literature, the information depths for our SRPES and XPS energies for TiC are approximately 1 nm and 2 nm, respectively [17].

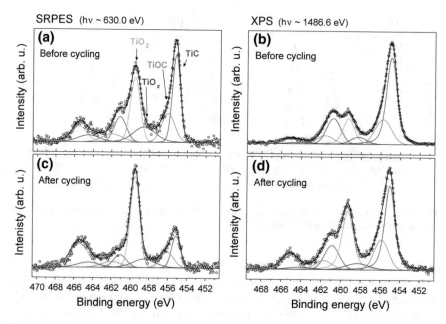

Fig. 3.40 SRPES and XPS spectra of Ti 2p before (**a, b**) and after (**c, d**) the electrochemical cycling (spectra are normalized to the maximum peak intensity) (reproduces from [14] with permission from Elsevier)

It can be seen, that even before the aging procedure Ti within the support material is not present solely in the TiC chemical state but also in oxidic states such as TiOC, TiO_2 and TiO_x [19] (Fig. 3.40a, b). By comparing the XPS and SRPES spectra which differ in information depth one can conclude that the titanium oxide and oxycarbide are relatively more abundant in the near-surface regions. Spectra measured after the aging procedure reveal the increase of TiO_2 peaks intensity at the expense of carbidic species; this effect is again more pronounced in the near-surface area of the sample (Fig. 3.40c, d). The TiC chemical state, essential for electron conductivity, is however still present and clearly distinguishable in both near-surface (SRPES) and subsurface (XPS) regions. This experiment proves that TiC is capable of withstanding high anodic potentials which are inevitable during the OER. Although partial oxidation of the TiC is observed, it is predominantly a surface phenomenon, the extent of which is already markedly repressed in the close subsurface region. The above-mentioned leads to our conclusion that TiC is a suitable conductive support for anode electrocatalyst in PEM-WE applications.

To sum up all the findings of this subchapter we can conclude that an alternative method of the CCM preparation with supported anode electrocatalyst for PEM-WE applications was presented. In contrast to more conventional approaches, our support material consists solely of conductive TiC nanoparticles and Nafion ionomer with no dispersed catalyst within. The anode electrocatalyst is deposited onto the surface of support sublayer only after it was hot-pressed to PEM, resulting in formation of so-to-say localized TPB.

Parameters relevant to PEM-WE performance such as loading of support material forming the sublayer, ionomer content within the support sublayer and amount of the Ir electrocatalyst on the surface of the support sublayer were all systematically varied and their influence was investigated. Ideal parameters regarding the support sublayer were found out to be 0.1 mg cm^{-2} of the support material which contained 15 wt% of Nafion ionomer in respect to mass of TiC nanoparticles. Using magnetron sputtering allowed us to coat the surface of optimized support sublayer with Ir thin films of a very low loadings. We obtained remarkable PEM-WE performance, both in terms of absolute current and also considering the amount of used Ir electrocatalyst.

Electrochemical cycling in conjunction with XPS analysis showed that sputtered Ir thin film, which was mainly in metallic state, underwent full oxidation to IrO$_2$ when exposed to high anodic potentials. Accelerated aging test also proved sufficient corrosion resistance of the anode support material. Although moderate oxidation of TiC was noticed, combination of photoelectron techniques with different information depth revealed that the effect is predominantly superficial and electrically conductive TiC state is still present. This renders investigated support material suitable for the role of conductive anode catalyst support in PEM-WE applications.

3.3.5   Further Optimization of Experimental PEM-WE MEA with Ir Thin-Film Catalyst Supported on TiC Particles (PEM and Anode GDL)

Up to this point, we have been optimizing the preparation process and composition of TiC-supported Ir thin-film catalyst for OER on anode side of PEM-WE. Obviously, there are plenty of other parameter that have significant effect on the performance of the cell. We are not referring to the operational temperature or water pressure but to the integral parts of the cell assembly such as PEM and anode GDL. We have been using Nafion N 115 (127 μm thick) which is well known golden standard for PEM-WE applications but there is wide range of other much thinner variants. As we have shown previously, the overall ohmic cell resistance is to large extent given by the PEM itself which is responsible for the transfer of protons from anode to cathode. The thinner the PEM, the higher the proton conductivity of the MEA and therefore also its performance. However, using thinner membrane means less mechanical stability (pressure resistance) and more prominent crossover effects [20]. Keeping this in mind, we decided to find the optimal PEM for our experimental PEM-WE setup. Membranes as thin as 28 μm (Nafion XL) are routinely used in PEM-FC R&D. In our PEM-WE cell though, such thin membrane did not survive even the initial break-in procedure. Presumably the pressure of water, pushed by peristaltic pump, was too high for the membrane to hold. Similarly, Nafion NR 212 (51 μm) ruptured after several minutes of testing. In Nafion NE 1035 (89 μm) we finally found the ideal combination of sufficient mechanical stability and considerably lower thickness in comparison to previously used N 115. Nafion NE 1035 withstood full scale experimental testing for more than four days and did not show any signs of depreciation.

Next we searched the market for more porous alternative to our standard Ti mesh which we used as anode current collector/GDL. It turned out that options are rather limited. Although suppliers of components for PEM-FC and PEM-WE offer so-to-say Ti felts with higher specific surface (see Fig. 3.9), the thickness of such felt was too high for our application (i.e. we could not properly seal the cell). Therefore, we had to look outside the sector of PEM technologies. Eventually we found the right product in portfolio of company Mott specializing in filtration and flow control for fluids in aerospace industry. We ordered sheets of sintered porous Ti filter with the same thickness to that of our standard Ti meshes. Detail on much denser and more porous structure of newly obtained current collector/GDL compared to our standard Ti mesh can be found in Fig. 3.41.

Similarly to Ti mesh, also the surface of sintered porous Ti GDL has to be Pt-treated to prevent the formation of nonconductive TiO_2.

Performance comparison of MEA with standard N 115 PEM (127 μm) + Ti mesh GDL and MEA with thinner NE 1035 PEM (89 μm) + sintered porous Ti GDL after break-in procedure can be seen in Fig. 3.43. Full details of both MEAs are summed up in Table 3.8 and are for better comprehension visualized in Fig. 3.42.

It can be seen that using thinner membrane and substituting Ti mesh with much more porous sintered Ti GDL yielded remarkably higher performance. For the first time and by a large margin we surpassed the benchmark of PEM-WE efficiency (black dots), which we set at the beginning of the Sect. 3.3 in Fig. 3.7, while having

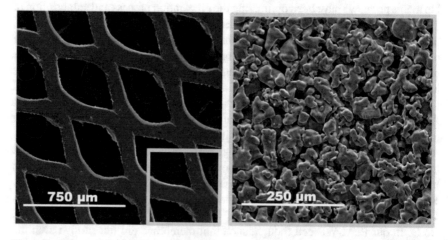

Fig. 3.41 Left—our previously used Ti mesh; right—new sintered porous Ti GDL (the square in left picture has the same view field as the picture on the right)

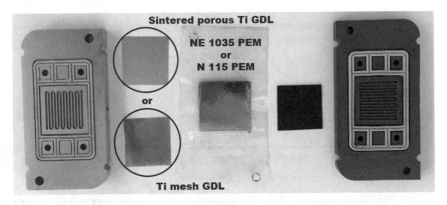

Fig. 3.42 MEAs—Ir supported on TiC "top", PEM and anode GDL optimization

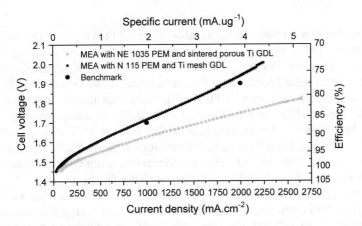

Fig. 3.43 IV curves of MEAs—**Ir supported on TiC "top", PEM and anode GDL optimization**

Table 3.8 Details of MEAs—**Ir supported on TiC "top", PEM and anode GDL optimization**

Cell	TiN-coated Ti anode end plate, graphite cathode end plate (active area 4.4 cm^2)
Membrane	**Nafion N 115 (127 μm thick) or Nafion NE 1035 (89 μm)**
Anode side	50 nm of Ir (113 μg cm^{-2}) sputtered on TiC-based sublayer 0.1 mg cm^{-2} (15 wt% of ionomer), hot-pressed on anode side of the membrane **Ti mesh GDL (FuelCellsEtc) + 50 nm of Pt sputtered on top of it** or **Sintered porous Ti GDL (Mott Corp.) + 50 nm of Pt sputtered on top of it**
Cathode side	Commercial GDE with 0.4 mg cm^{-2} of Pt (Alfa Aesar)

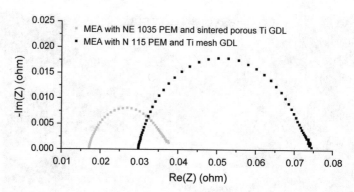

Fig. 3.44 Nyquist diagrams of **MEAs—Ir supported on TiC "top", PEM and anode GDL optimization** obtained at 1.5 V

significantly lower loading of noble metal on anode (just 113 µg cm^{-2}) in comparison to standards PEM-WEs.

By analyzing data obtained by PEIS at 1.5 V it is clearly confirmed that the performance boost indeed comes from the combination of two factors (see Fig. 3.44). Firstly, the ohmic resistance of MEA with thinner membrane is approximately 0.013 Ω lower (0.017 Ω); this is mainly associated with lower proton resistance due to the thinner membrane. Secondly, smaller semi arc of Nyquist plot of MEA with more porous sintered Ti sheet from Mott suggests higher reaction kinetics which is in agreement with the idea of higher specific surface of this GDL and hence better, more even distribution of electric potential over the Ir thin film.

Interestingly enough, when we tried to further increase the surface of sintered porous Ti GDL by spraying the mixture of TiC and Nafion on top of it (total loading of 0.1 mg cm^{-2}, 15 wt% Nafion),[13] the performance did not improve; on the contrary it worsened slightly. SEM micrograph of sintered porous Ti GDL with additional TiC-based layer on top of it is in Fig. 3.45. The performance comparison of **MEA with plane sintered porous Ti GDL** (the superior curve from Fig. 3.43) versus **MEA with sintered porous Ti GDL + TiC-based layer on top of it** after the break-in procedure is shown in Fig. 3.47. **MEA with sintered porous Ti GDL + TiC-based layer on top of it** is shown in Fig. 3.46 and its details are summed up in Table 3.9.

Nyquist diagrams obtained by PEIS at 1.5 V suggest that lower performance of **MEA with sintered porous Ti GDL + TiC-based layer on top of it** is given by both higher ohmic resistance and slower kinetics (see Fig. 3.48). Admittedly, the addition of TiC nanoparticles with the aim of increasing the active surface of GDL did not have the desired effect of improving the PEM-WE efficiency. Apparently, having another layer of TiC-based material, albeit on the other side of the Ir catalyst, causes a rise in the aforementioned overall ohmic resistance of the MEA (not unlike in Fig. 3.29). Regarding the slower kinetics, we are inclined to believe that addition

---

[13]The same loading and composition of TiC-based support that was also hot-pressed on anode side of the PEM and then sputtered over by Ir thin film.

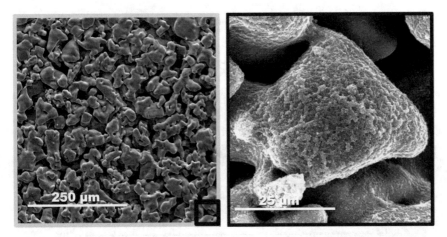

Fig. 3.45 Left—sintered porous Ti GDL; right—same sintered porous Ti GDL with additional 0.1 mg cm^{-2} of TiC-based layer on top of it (15 wt% of Nafion in respect to TiC) (the dark square in the left picture has the same view field as the picture on the right)

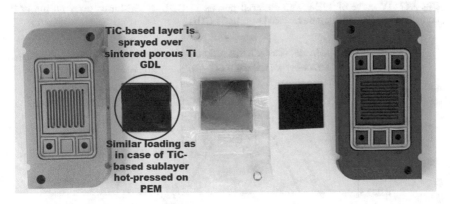

Fig. 3.46 MEA with sintered porous Ti GDL + TiC-based layer on top of it

Table 3.9 Details of **MEA with sintered porous Ti GDL + TiC-based layer on top of it**

Cell	TiN-coated Ti anode end plate, graphite cathode end plate (active area 4.4 cm^2)
Membrane	Nafion NE 1035 (89 μm)
Anode side	50 nm of Ir (113 μg cm^{-2}) sputtered on TiC-based sublayer 0.1 mg cm^{-2} (15 wt% of ionomer), hot-pressed on anode side of the membrane Sintered porous Ti GDL (Mott Corp.) + 50 nm of Pt sputtered on top of it + 0.1 mg cm^{-2} TiC-based layer (15 wt% Nafion)
Cathode side	Commercial GDE with 0.4 mg cm^{-2} of Pt (Alfa Aesar)

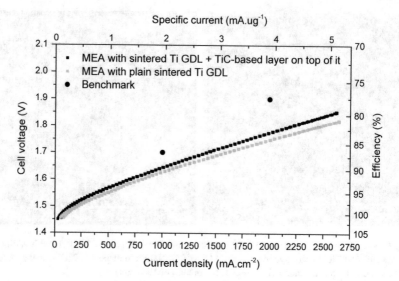

Fig. 3.47 IV curve of **MEA with sintered porous Ti GDL + TiC-based layer on top of it** compared to **MEA with plane sintered porous Ti GDL**

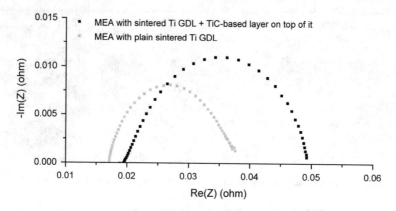

Fig. 3.48 Nyquist diagrams of **MEA with sintered porous Ti GDL + TiC-based layer on top of it** compared to **MEA with plane sintered porous Ti GDL** obtained at 1.5 V

of TiC-based layer on top of sintered porous Ti GDL not only leads to the desired higher surface on individual grains but also to partial clogging of pores which to certain extent hinders the mass transport of water and/or oxygen bubbles between flow field and CCM.

To sum up the results of Sect. 3.3.4 we can conclude that using a thinner membrane and denser anode GDL, while keeping all other relevant parameters unaltered, expectedly increased the performance of PEM-WE. Substituing standard Ti mesh with unique denser sintered porous Ti GDL and going down from 127 μm of standard Nafion N 115 to 89 μm of more modern Nafion NE 1035 led to an efficiency rise of about 5% at 1 A cm^{-2} and approximately 10% at 2 A cm^{-2}. However, further thinning of the PEM to dimensions more typical for PEM-FC (51 μm of Nafion NR 212 or 28 μm of Nafion XL) resulted in mechanical instability; the membrane ruptured eventually. Also, the attempt to further increase the active area of sintered porous Ti GDL by spraying it over with TiC nanoparticles did not yield better performance in PEM-WE cell. Apparently, the positive effect of higher surface was significantly suppressed by undesired decrease in pore permeability.

Even though thinning of the PEM and increasing of the anode GDL surface clearly has its limits from the performance point of view, the use of NE 1035 and unmodified variant of unique porous sintered Ti GDL alone leads to the most important outcome of this subchapter. We finaly surpassed the PEM-WE efficiency benchmark (Fig. 3.7) in spite of using just a fraction of anode catalyst loading in comparison to common MEAs. The ambitious goal we set out at the beginning of Chap. 3 was thus fulfilled.

Also, considering the results of the past three subchapters, it is now safe to say that we successfully proved the convenience of using thin-film magnetron sputtering for low-loading PEM-WE anode catalyst deposition. Hence, we declare that the second objective of the thesis has been completed.

3.4 Thin-Film Magnetron Sputtered Catalyst for PEM-URFC

After accomplishing our goal of preparing efficient, thin-film, low-loading anode catalyst for PEM-WE, we investigated the possibilities of using magnetron sputtered catalyst for PEM-URFCs as well.[14] As is exhaustively described in Sect. 1.4 there are two possible configurations of URFCs; the oxidation and reduction electrode configuration and the H$_2$ and O$_2$ electrode configuration. We decided to set our URFC cell in the first arrangement which keeps the redox roles of the electrodes unchanged. As such OER and HOR were paired on the anode and HER and ORR on the cathode. In this way, the faster reaction of PEM-WE is paired with the slower reaction of PEM-FC and vice versa. Since we were building on our results of the previous chapter, we planned to modify the promising Ir/TiC OER catalyst by adding Pt in order to make it also active towards HOR. The cathode catalyst was, similarly to experiments in previous section, kept referential; as both HER and ORR are catalyzed by platinum.

Before we proceed to discussing actual results of our experimental endeavor, it will be recapped what performances were we able to obtain for individual dedicated electrochemical devices, the PEM-WEs and PEM-FCs.

---

[14]Certain formulations and figures within Sect. 3.4 are taken from the submitted paper written by Kúš et al. [21].

3.4.1 Reference Performances of Dedicated PEM-WE and PEM-FC Cells

Unfortunately, there is no reliable purchasable referential anode catalyst for PEM-WE, nonetheless the performance we were able to obtain with Ir/TiC catalyst, after the optimization steps mentioned in Sects. 3.3.4 and 3.3.5, is well within the range of the state-of-the-art efficiencies. That being said, we will consider this MEA (see details in Table 3.10) to be our reference for PEM-WE performance (Fig. 3.49).

Regarding PEM-FC, the situation is simpler. Since Pt catalyzes both ORR and HOR, conventional GDE (i.e. carbon paper with a MPL of carbon black mixed with Pt nanoparticles on top of it) can serve as a state-of-the-art reference for both half-reactions. Performances of PEM-FC MEAs having the same reference catalysts but differing in used PEMs are plotted in Fig. 3.51. All relevant details of tested MEAs are for better clarity summarized in Table 3.11 and shown in Fig. 3.50. Unlike in

Table 3.10 Details of **MEA—PEM-WE reference for PEM-URFC**

Cell	TiN-coated Ti anode end plate, graphite cathode end plate (active area 4.4 cm^2)
Membrane	Nafion NE 1035 (89 μm)
Anode side	50 nm of Ir (113 μg cm^{-2}) sputtered on TiC-based sublayer 0.2 mg cm^{-2} (15 wt% of ionomer)[a], hot-pressed on anode side of the membrane Sintered porous Ti GDL (Mott Corp.) + 50 nm of Pt sputtered on top of it
Cathode side	Commercial GDE with 0.4 mg cm^{-2} of Pt (Alfa Aesar)

[a]Although using 0.1 mg cm^{-2} of TiC-based sublayer leads to better performances, we decided to use higher loading for URFC experiments because the consistency of flawless hot pressing was much higher

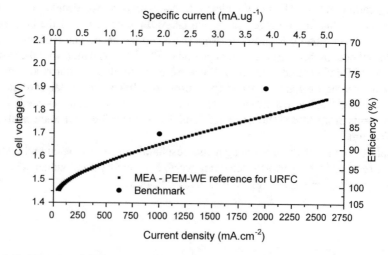

Fig. 3.49 IV curves of **MEA—PEM-WE reference for PEM-URFC**

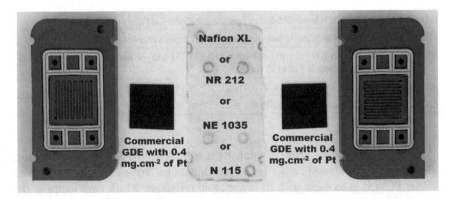

Fig. 3.50 MEAs—**PEM-FC reference for PEM-URFC**; membranes with different thickness were tested

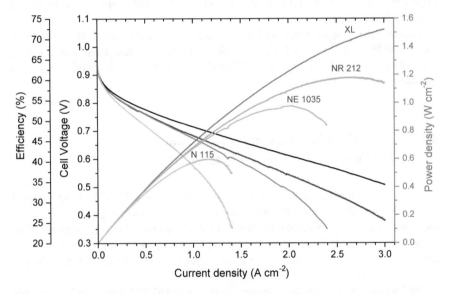

Fig. 3.51 IV (black/grey) and power (green) curves of H_2/O_2 PEM-FCs with a standard state-of-the-art Pt/C anode and cathode for different Nafion membrane thicknesses (reproduces from [22] with permission from Elsevier) (black curves are related to the right Y-axis (Efficiency and Cell voltage), green curves to the right Y-axis (Power density); X-axis is shared for both sets of curves)

case of PEM-WE, the PEM-FC cell was kept at 70 °C (the efficiencies are calculated accordingly), the press cell pressure was kept the same (8 bar). The temperature of O_2 and H_2 humidifiers was held at 65 °C. The flow rates of hydrogen and oxygen were set to 40 and 30 sccm, respectively,[15] against 1.5 bar of backpressure.

---

[15]These parameters are kept the same for later measurements of PEM-URFC in FC mode.

Table 3.11 Details of **MEA—PEM-FC reference for PEM-URFC**

Cell	Graphite cathode and anode end plates (active area 4.4 cm^2)
Membrane	**Nafion XL (28 μm)** or **Nafion NR 212 (51 μm)** or **Nafion NE 1035 (89 μm)** or **Nafion N 115 (127 μm)**
Anode side	Commercial GDE with 0.4 mg cm^{-2} of Pt (Alfa Aesar)
Cathode side	Commercial GDE with 0.4 mg cm^{-2} of Pt (Alfa Aesar)

It can be clearly seen that the thinner the membrane the higher the efficiency. As was already discussed earlier in the text, higher performance of the MEA with thinner PEM is associated with its lower ionic resistance which contributes to the overall ohmic resistance of the cell. Due to the fact that PEM-URFC needs to operate in both regimes, the relevant reference for us is the curve of MEA with NE 1035 PEM which is the thinnest membrane, we have tested, that is stable also under the PEM-WE conditions.

Furthermore, it should be noted that the same commercial catalyst was used on cathode of PEM-WE and on both electrodes of PEM-FC. This proves that it can be conveniently used as a bifunctional ORR/HER catalyst in PEM-URFC (i.e. on the cathode).

3.4.2 Thin-Film Bifunctional Anode Catalyst for PEM-URFC (Pt–Ir Co-sputtering)

The most straightforward approach to prepare a bifunctional thin-film anode catalyst is to simply introduce a second sputtering source to the deposition process of catalyst onto the TiC-based support sublayer (see Fig. 3.52). Following this idea, we calibrated the sputtering rate of Pt to match the one of Ir and by means of co-sputtering we consequently prepared a 100 nm thin mixed layer (50 nm of Ir together with 50 nm of Pt). Details of the MEA are in Table 3.12.

We had already proven that the surface of optimized TiC-based sublayer is able to support catalytic layer up to 75 nm thick without any significant morphological change or loss in roughness of the surface (see Fig. 3.34d). The addition of extra

Table 3.12 Details of **MEA—bifunctional co-sputtered Pt–Ir catalyst**

Cell	TiN-coated Ti anode end plate, graphite cathode end plate (active area 4.4 cm^2)
Membrane	Nafion NE 1035 (89 μm)
Anode side	100 nm of Pt–Ir (220 μg cm^{-2}) co-sputtered on TiC-based sublayer 0.2 mg cm^{-2} (15 wt% of ionomer), hot-pressed on anode side of the membrane Sintered porous Ti GDL (Mott Corp.) + 50 nm of Pt sputtered on top of it
Cathode side	Commercial GDE with 0.4 mg cm^{-2} of Pt (Alfa Aesar)

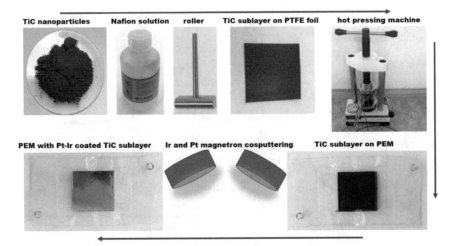

Fig. 3.52 Schematic representation of CCM preparation with TiC-supported thin-film Pt–Ir anode bifunctional catalyst ("top" geometry)

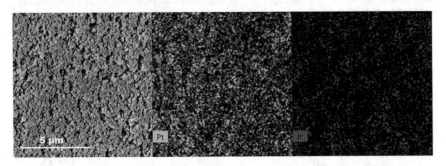

Fig. 3.53 SEM image of 0.1 mg cm^{-2} TiC-based support sublayer with 100 nm of Pt–Ir catalyst sputtered on top of it and EDX mapping of Pt and Ir (same magnification)

25 nm did not result in any noticeable difference (see Fig. 3.53a). The EDX mapping was again used for verification of homogeneous distribution of Ir and Pt (Fig. 3.53b, c) as well as for confirmation of Pt:Ir ratio using standardless PB-ZAF quantification. The determined ratio was 50:50 ± 5%.

Interestingly enough, upon the examination of 10 nm thin Pt–Ir layer in XPS (spectra were corrected to charge shift setting the C 1s spectra to 284.5 eV), we found out that co-sputtering of Pt and Ir most likely led to formation of Pt–Ir alloy. This can be concluded based on apparent Pt binding energy upshift and Ir downshift relative to metallic Pt and Ir reference (sputtered individually on the same type of substrate—Ti foil) (Fig. 3.54) [23]. Such binding energy shifts can be correlated with intra-atomic charge transfer from Pt to Ir [24].

The PEM-WE in cell performance after the break-in procedure can be seen in Fig. 3.55. The obtained IV curve is compared to the referential curve from Fig. 3.49.

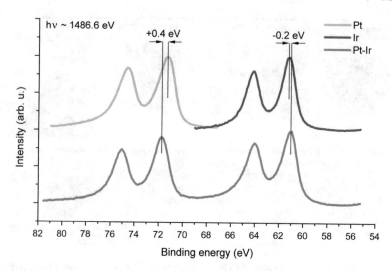

Fig. 3.54 XPS spectra of Ir 4f and Pt 4f of pure Pt and pure Ir on Ti foil as well as Pt–Ir on the same type of substrate; spectra are normalized to the maximum peak intensity and have subtracted Shirley type background

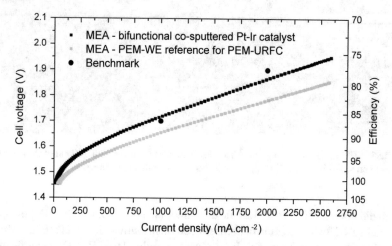

Fig. 3.55 IV curve of **MEA—bifunctional co-sputtered Pt–Ir catalyst** compared to **MEA—PEM-WE reference for PEM-URFC**

MEA with bifunctional anode catalyst is clearly showing lower performance in comparison to MEA with pure Ir reference. It does not even meet the benchmark of 1 A cm^{-2} at 1.7 V. PEIS measurements (Fig. 3.56) confirm that performance loss is due to the reaction's kinetics, not the ohmic resistance. In other words, the addition of 50 nm of Pt does not alter the overall MEA conductivity in any significant way but does somehow hinder the ability of Ir to catalyze OER.

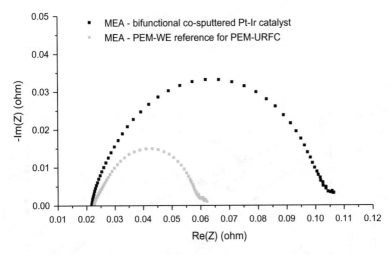

Fig. 3.56 Nyquist diagrams of **MEA—bifunctional co-sputtered Pt–Ir catalyst** compared to **MEA—PEM-WE reference for PEM-URFC**

After the performance test in PEM-WE regime, we proceeded to the PEM-FC operation. The water circuit and H_2 output tube (see Fig. 3.5) were disconnected and the press cell itself was transferred to the dedicated PEM-FC testing station.[16]

It should be emphasized that although the temperature was not controlled during the transfer (heating cartridges were switched off), the press cell pressure was kept constant at 8 bar. Once the gases inlets of PEM-FC testing station were attached to the press cell and its temperature was set to 70 °C, the anode and cathode sides of the experimental cell were flushed with dry nitrogen in order to get rid of residual H_2, O_2 and excess humidity. The temperature of humidifiers and individual gas flow was similar to the previous referential experiment (see Sect. 3.4.1). IV and power curves obtained after brief initiation period (10 min constant voltage of 0.6 V), compared to the reference curves from Fig. 3.51 can be seen in Fig. 3.57.

MEA-bifunctional co-sputtered Pt–Ir catalyst yielded considerably lower peak performance than MEA–PEM-FC reference for PEM-URFC; 0.48 W cm^{-2} at 27.5% efficiency in comparison to 0.98 W cm^{-2} at 32.5%. Such significant decrease is due to several reasons. Firstly and most importantly, the anode Pt catalyst was in case of referential MEA supported on commercial carbon black in form of MPL, carried on carbon paper GDL. The bifucntional Pt–Ir catalys, on the other hand, was sputtered over TiC-based sublayer which was overlaid by porous sintered Ti GDL. Although TiC-based sublayer in conjunction with porous sintered GDL features relatively high surface utilization, it is still less effective than dedicated fuel cell carbon black/carbon paper configuration. Pure carbon however cannot be present on OER side of the cell so there is no potential in improvement here. Secondly, loading of the thin-film Pt catalyst on the bifunctional MEA is roughly one fourth of that on a referential MEA.

---

[16]More details on PEM-FC testing setup can be found in [25].

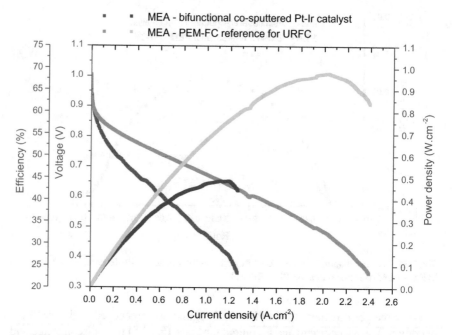

Fig. 3.57 IV and power curves (the IV curves are in tones of red and are coupled with left Y-axises, the power curves are in tones of blue and are coupled with right Y-axis, the X-axis is shared with both sets of curves) of **MEA—bifunctional co-sputtered Pt–Ir catalyst** compared to the **MEA—PEM-WE reference for PEM-URFC**

Finally, it has to be stressed that the anode back-plate used with bifunctional MEA was TiN-coated Ti whereas referential MEA was coupled with the standard carbon one. Carbon has better conductivity than TiN, but again, it cannot be part of PEM-URFC due to the aggressive nature of OER potential. Overall, we can conclude that because of above mentioned factors, certain loss in efficiency was simply inevitable.

After the testing of **MEA—bifunctional co-sputtered Pt–Ir catalyst** in PEM-FC mode, the cell was reconnected to the PEM-WE testing setup. After stabilization of the temperature at 80 °C and brief rehumidification period the performance of bifunctional MEA was reexamined in PEM-WE regime. Curve similar to that in Fig. 3.55 was obtained, proving that **MEA—bifunctional co-sputtered Pt–Ir catalyst** repeatedly works in both operational regimes without any noticeable efficiency deterioration.

Chemical analysis

IV curves shown in previous section proved that 100 nm Pt–Ir thin film prepared by multitarget magnetron co-sputtering, can be used as a low-loading anode catalyst in PEM-URFC. Performance of tested bifunctional MEA in both regimes was however inferior to that of dedicated referential MEAs in individual PEM-FC and PEM-WE modes. While a significant drop in PEM-FC efficiency of bifunctional MEA

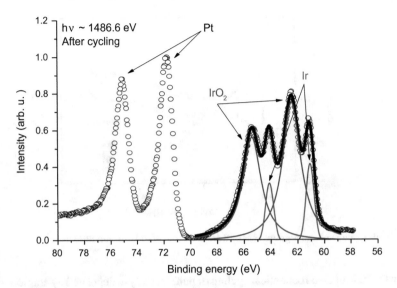

Fig. 3.58 XPS spectra of Pt 4f and Ir 4f after the electrochemical cycling; spectra are normalized to the maximum peak intensity

was somehow expected, since high-surface/high conductivity carbon-made materials cannot be used in PEM-URFC regardless of what sort of catalyst is actually used on the anode, the reason for lower efficiency in PEM-WE regime is more unclear. After all, the reference MEA and bifunctional MEA were identical in terms of GDL, even the catalyst loading of iridium was the same. In order to shed more light on the Pt–Ir bifunctional co-sputtered thin-film catalyst we decided to further explore its chemical behavior. Initial XPS analysis (Fig. 3.54) hinted to alloy formation, the question therefore arises whether such alloy reacts differently to the applied potential in comparison to pure catalyst. Similarly to the experiments done on TiC-based support sublayer and pure Ir catalyst, also with the Pt–Ir co-sputtered catalyst[17] we conducted 500 potential cycles in range from 0 to 1.7 V versus RHE at 50 mV s^{-1} in 0.1 M H$_2$SO$_4$ aqueous solution (last scan was terminated at 1.7 V). A sample was then transferred through inert N$_2$ atmosphere to the XPS apparatus for post cycling analysis. Charge-corrected (to C 1s line at 284.5 eV) Pt 4f and Ir 4f spectra acquired after the electrochemical cycling with substracted Shirley-type background can be seen in Fig. 3.58.

While Pt 4f doublet seems to be unaltered with Pt 4f 7/2 peak at the exactly same binding energy of 71.8 eV as before the electrochemical cycling (Fig. 3.54), the Ir spectra changed significantly. It is expected to see oxidation of metallic Ir, after all, as stated previously the IrO$_2$ is the actual catalyst of OER and its presence is needed. What is interesting though is that the oxidation in Fig. 3.58 is only partial; Ir^0 metallic state is still undoubtedly present. This is a fundamentally different outcome

---

[17]It was the very same sample which was measured in Fig. 53, i.e. 10 nm Pt–Ir on Ti foil.

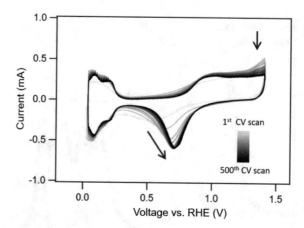

Fig. 3.59 Cyclic voltammograms obtained from co-sputtered Pt–Ir thin film

as in the case of electrochemical cycling of pure Ir catalyst, deposited by single-target sputtering, which lead to full oxidation (Fig. 3.39). These results seem to suggest the incomplete oxidation of Ir, in other words its insufficient electrochemically-induced activation, is directly responsible for lower OER activity in comparison to the reference (Fig. 3.55).

In order to decipher what is the actual reason for incomplete oxidation of Ir within the Pt–Ir alloy, we studied it further by means of electrochemical atomic force microscopy (EC-AFM). This method, as described in the Experimental section, allows to perform three-electrode electrochemical measurements on the sample and to quasi-simultaneously investigate its potential-driven topography transformations while being directly in the liquid environment of the electrolyte (see Fig. 2.5). For the experiment we sputtered 10 nm of Pt–Ir alloy on freshly cleaved highly oriented pyrolytic graphite (HOPG).[18] A sample was put to the dedicated EC-AFM cell, filled with 0.1 M H_2SO_4 electrolyte. The surface of Pt–Ir thin film was probed using Bruker ScanAsyst Fluid + tips in tapping mode. The 500×500 nm^2 images were obtained before the electrochemical cycling and then after completion of 50, 100, 150, 200 and 400 cycles in range from 0 to 1.4 V versus RHE[19] with 200 mV s^{-1} scan rate. Cyclic voltammograms are plotted in Fig. 3.59; individual AFM images are shown in Fig. 3.60.

Interestingly enough, already after the first initial cycles, which can be considered as cleaning sweeps, the cyclic voltammogram of Pt–Ir thin film stabilized itself in a shape practically indistinguishable from the typical voltammogram of pure sputtered Pt thin film [26]. One can identify the adsorbtion/desorbtion of hydrogen in the lower potential region, typical for metallic Pt. The onset of platinum oxidation occurring at about 0.8 V in the anodic scan as well as consequent complete reduction of formed

---

[18] HOPG is a highly pure and ordered form of synthetic graphite with well aligned individual laterally oriented crystallites.

[19] Again, miniature leak-free Ag/AgCl (3.4 M KCl) electrode was used as reference; Pt wire as counter electrode.

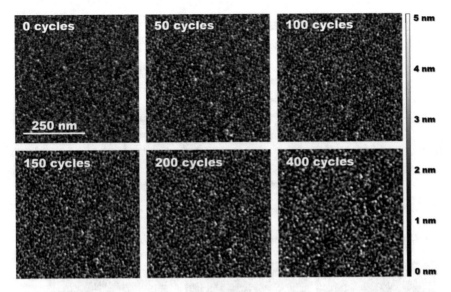

Fig. 3.60 AFM images of 10 nm Pt–Ir thin film after certain number of electrochemical cycles in range 0 V–1.4 V_{RHE}; all images are taken with the same magnification and same height contrast

oxides back to metallic Pt in cathodic scan are also clearly visible. However, no extra humps or peaks which could be associated with iridium oxidation process were observed. In fact, not a single clue even hinted the very presence of Ir within the layer. Looking solely at the stabilized cyclic voltammogram itself, one could easily assume that the sputtered thin layer contains polycrystalline Pt only.

The AFM images which tracked the topographical changes of Pt–Ir thin film show very mild coarsening of its surface. It could be associated with dissolution and redeposition of material at higher potentials, based on the voltammogram most likely Pt, leading to the growth of larger nanoparticles due to the Ostwald ripening effect [27, 28].

Following the EC-AFM measurement on Pt–Ir thin film we conducted similar experiment with 10 nm Ir thin film in order to confirm its distinct behavior when present in pure non-alloyed form. Indeed, cyclic voltammograms look completely different (Fig. 3.61). In anodic scan, at potentials above 0.45 V commences the reversible oxidation of Ir, which leads to formation of a compact Ir oxide film. However unlike in case of Pt (or Pt–Ir respectively), when the potential is repeatedly cycled to potentials higher than 1.3 V, Ir is also gradually oxidized in an irreversible manner [29]. This is evident by progressive broadening of the voltammograms (larger currents) with increasing number of underwent cycles, reflecting continuous increase in active area of the formed oxide through place-exchange mechanism [30, 31].

Correspondingly, AFM images reveal significant topographical changes of thin film's surface as it repeatedly transitions from reduced to oxidized state (see Fig. 3.62). Coarsening and roughening of a much higher degree than in case of

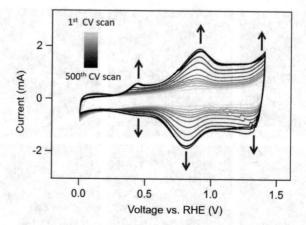

Fig. 3.61 Cyclic voltammograms obtained from pure sputtered Ir thin film

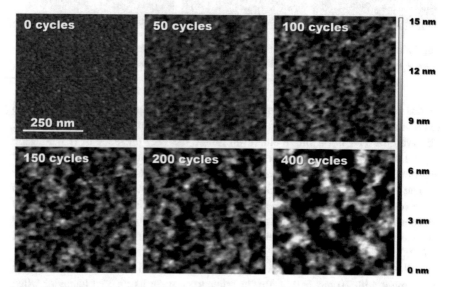

Fig. 3.62 AFM images of 10 nm Ir thin film after certain number of electrochemical cycles in range 0 V–1.4 V_{RHE}; all images are taken with the same magnification and same height contrast

Pt–Ir thin film can be observed, due to replacement of Ir by the larger volume of IrO_x material [32].

Summing up data from XPS (Figs. 3.39, 3.54, 3.58) and EC-AFM analysis (Figs. 3.59, 3.60, 3.61, 3.62) from both Pt–Ir and pure Ir thin films and correlating them with corresponding PEM-WE performances (Fig. 3.55) we believe that upon co-sputtering of Pt and Ir, or possibly shortly after introduction of high anodic potential, the thin film rearranges itself into the Ir-rich core/Pt-rich shell configuration rather than staying in form of simple homogenous alloy. As such, unlike in case

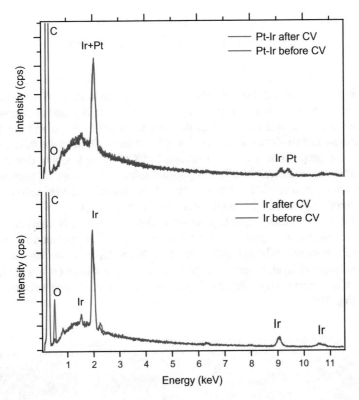

Fig. 3.63 EDX analysis of 10 nm Pt–Ir (top) and 10 nm Ir (bottom) thin films before and after cyclic voltammetry measurements

of pure Ir, the Ir within Pt–Ir thin film is unable to undergo full oxidation necessary for efficient OER activity and thus performance of this catalyst in PEM-WE mode is insufficient. Admittedly, the chemical state of Ir^{4+} is partially present, as revealed by XPS, but clearly most of the thin film's volume, including it subsurface region, contains metallic Ir which is further backed up by more bulk-sensitive EDX analysis (see Fig. 3.63). Spectra of Pt–Ir remain relatively unaltered after the cyclic voltammetry while spectra of pure Ir thin film feature distinct peak of oxygen reflecting its thorough oxidation.

Based on the aforementioned results we drew a conclusion that in order to achieve efficient performance of thin-film Pt–Ir bifunctional catalyst in both operational regimes of PEM-URFC (i.e. PEM-WE and PEM-FC) metallic Pt and oxidic IrO_2 states have to be present. We have proven, that co-sputtering of Ir and Pt in Ar atmosphere is not the ideal way to achieve this.

3.4.3 Thin-Film Bifunctional Anode Catalyst for PEM-URFC (Pt, Ir Sandwich Sputtering)

The previous chapter dealt with co-sputtered Pt–Ir thin films which met the expectations in PEM-FC mode, however they underperformed in PEM-WE mode. Insufficient extent of Ir oxidation within Pt–Ir thin film, possibly due to formation of core-shell structure, was identified as the reason behind it. We found out that for Ir to undergo full oxidation it has to be sputtered alone. To achieve this, we again modified the preparation process of reversible CCM in such a way that Pt and Ir thin film are sputtered individually on opposing sides of TiC-based support sublayer, creating so-to-say a sandwich-like structure. It is basically a combination of "top" and "bottom" approaches thoroughly described in Sect. 3.3.4.

The ink of TiC-based support sublayer is spread over PTFE transient foil and let dry. Afterwards it is sputtered over by Pt and hot-pressed on Nafion membrane, forming Pt "bottom" HOR catalytic layer. Finally, the top, still uncoated surface of TiC-based support layer, recently hot-pressed on PEM, is coated by Ir. For clarification, the whole process is depicted in Fig. 3.64, while the CCM cross-section can be seen in Fig. 3.65.

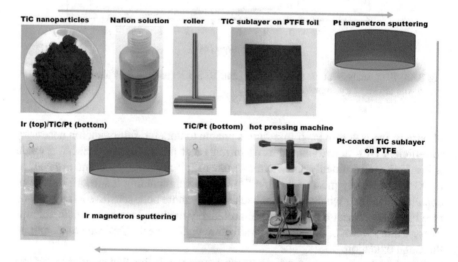

Fig. 3.64 Schematic representation of CCM preparation with TiC-supported sandwich-like thin-film Ir (top), Pt (bottom) anode catalyst for PEM-URFC

Fig. 3.65 Schematic cross-section of thin-film sandwich-like TiC-supported Ir (top), Pt (bottom) anode catalyst for PEM-URFC

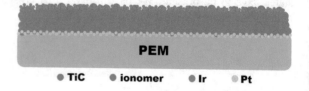

Our previous results from Sect. 3.3.4 suggested that the "bottom" configuration is not working properly in case of OER on PEM-WE anode due to poor mass transport of water to the localized TPB and thus reagent starvation of Ir catalyst buried under the TiC-based sublayer. In addition, produced O_2 bubbles might stay encapsulated under the support layer, thus ending up blocking the active sites of the catalyst. This however should not apply for HOR on PEM-FC anode catalyzed by Pt.

After all, molecule of H_2 is much smaller than H_2O but most importantly hydrogen is introduced in gaseous state not liquid; its permeation through TiC-based sublayer down to Pt catalyst should be therefore sufficient. Moreover, both intermediate products of anode PEM-FC half-reaction (proton and electron) are dragged to the cathode side of the cell, unlike in case of OER and O_2 bubbles; as such blocking of Pt active sites should be minimal.

To prove our assumptions, we prepared bifunctional MEA with sandwich-like geometry; details of which can be seen in Table 3.13.

The PEM-WE in cell performance of bifunctional sandwich-like MEA after the break-in procedure is plotted in Fig. 3.66. The obtained IV curve is compared to the curve of MEA—PEM-WE reference for PEM-URFC as well as to the curve of MEA—bifunctional co-sputtered Pt–Ir catalyst.

It is immediately clear that the sandwich-like thin-film catalyst performed much better than its co-sputtered counterpart. In fact, the MEA with sandwich-like structure exhibited virtually the same efficiency as the reference MEA. This hints that the presence of Pt thin film located under the TiC-based sublayer in position "bottom" does not hinder the performance of Ir catalyst in PEM-WE mode in any observable way. Indeed, the PEIS analysis (Fig. 3.67) confirms that the overall ohmic cell resistance (intercept of the X-axis) of sandwich-like MEA is indistinguishable from the referential one. Also, the semi arcs within the Nyquist plot sit more or less on each other proving basically the same reaction kinetics.

It is fair to note that in terms of specific performance (per loading of noble metals), the relative difference in efficiency of both bifunctional MEAs in respect to the reference MEA is more prominent (see Fig. 3.68). This is obviously due to the higher total loading of bifunctional MEAs, more specifically the addition of 50 nm of Pt (107 $\mu g\ cm^{-2}$) to the anode side of PEM.

After the performance test in PEM-WE regime, we proceeded to the investigation of PEM-FC operation. Similarly to the previous experiment, the water circuit and

Table 3.13 Details of **MEA—bifunctional sandwich-like Pt (bottom) Ir (top) catalyst**

Cell	TiN-coated Ti anode end plate, graphite cathode end plate (active area 4.4 cm^2)
Membrane	Nafion NE 1035 (89 μm)
Anode side	50 nm Pt sputtered on TiC-based sublayer 0.2 mg cm^{-2} (15 wt% of ionomer) then hot-pressed on anode side of the membrane and sputtered over by 50 nm Ir, total loading Pt + Ir (220 $\mu g\ cm^{-2}$) Sintered porous Ti GDL (Mott Corp.) + 50 nm of Pt sputtered on top of it
Cathode side	Commercial GDE with 0.4 mg cm^{-2} of Pt (Alfa Aesar)

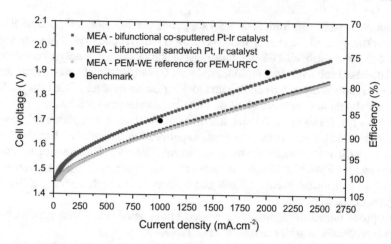

Fig. 3.66 IV curve of **MEA—bifunctional sandwich-like Pt (bottom) Ir (top) catalyst** compared to **MEA—bifunctional co-sputtered Pt–Ir catalyst** and **MEA—PEM-WE reference for PEM-URFC**

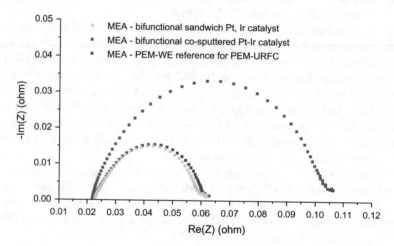

Fig. 3.67 Nyquist diagrams of **MEA—bifunctional sandwich-like Pt (bottom) Ir (top) catalyst** compared to **MEA—bifunctional co-sputtered Pt–Ir catalyst** and **MEA—PEM-WE reference for PEM-URFC**

hydrogen output tube were disconnected, heating cartridges switched off and the press cell itself (still holding 8 bars) was transferred to the dedicated PEM-FC testing station.

Once connected, the temperature was set to 70 °C and the anode and cathode sides of the experimental cell were flushed with dry nitrogen in order to get rid of residual H_2, O_2 and excess humidity. The temperature of humidifiers and individual

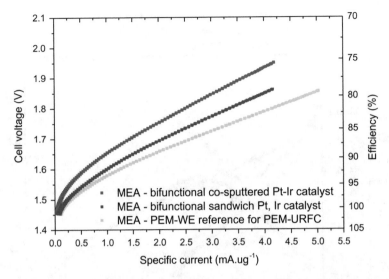

Fig. 3.68 Specific IV curves of **MEA—bifunctional sandwich-like Pt (bottom) Ir (top) catalyst, MEA—bifunctional co-sputtered Pt–Ir catalyst** and **MEA—PEM-WE reference for PEM-URFC**

gas flow was similar to the previous PEM-FC experiments (see Sect. 3.4.2) IV and power curves obtained after brief initiation period (10 min constant voltage of 0.6 V), compared to the results of co-sputtered and referential MEA from Fig. 3.57 are plotted in Fig. 3.69.

The measurement reveals that the sandwich-like bifunctional MEA was superior to the co-sputtered one also in the PEM-FC mode. Maximum power density of sandwich-like MEA surpassed 0.53 W cm^{-2} in comparison to 0.48 W cm^{-2} of co-sputtered MEA. Unlike in case of Ir, the chemical state of Pt within both MEAs is identical (the catalytically active metallic Pt^0), the rise in performance is thus most likely the result of thin film's positioning. Pt thin film within sandwich-like MEA is below the TiC-based support sublayer (i.e. hot-pressed in position "bottom" against the anode side of PEM), as such H$^+$ ions from HOR do not need to cross the sublayer, which would inevitably limit the ionic conductivity, but permeate more effectively straight through membrane (the effect was already described in Sect. 3.3.4).

When looking at the specific performances (per mass of noble metals) of individual MEAs, one can notice that the relative difference between maximum specific power of bifunctional MEAs and the referential one is smaller (see Fig. 3.70). This has to do with the fact that anode loading of our bifunctional thin-film MEAs is nearly two times lower.

Following the testing of **MEA—bifunctional sandwich-like Pt (bottom) Ir (top) catalyst** in PEM-FC mode, we reconnected the cell to the PEM-WE station to prove that the performance in PEM-WE mode remained the same. After stabilization of

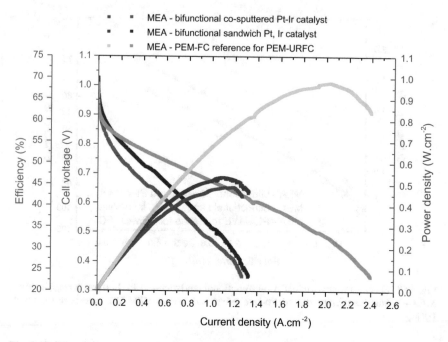

Fig. 3.69 IV and power curves of **MEA—bifunctional sandwich-like Pt (bottom) Ir (top) catalyst** compared to **MEA—bifunctional co-sputtered Pt–Ir catalyst** and **MEA—PEM-WE reference for PEM-URFC**

the temperature at 80 °C and rehumidification of the anode catalyst, we obtained performance curve similar to that in Fig. 3.66, confirming that tested MEA repeatedly works in both modes without deterioration.

Experiments in this chapter proved our hypothesis that separating the Ir and Pt thin films on the anode side of the PEM-URFC certainly leads to higher performance not only in PEM-WE but also in PEM-FC mode. We can now confidently say that thorough oxidation of Ir to IrO_2 (which was not possible when Ir was co-sputtered with Pt) is clearly responsible for higher efficiency in PEM-WE regime.

On the other hand, the less prominent increase of efficiency in PEM-FC regime seems to be due to the swapping of Pt thin film from position "top" to position "bottom" which allows for better ionic conductivity.

Summing up the outcomes of Sects. 3.4.2 and 3.4.3 we conclude that magnetron sputtering could indeed be conveniently used for deposition of thin-film, low-loading and efficient bifunctional catalysts for PEM-URFC. It is however crucial to carry out the sputtering in such a way that individual catalysts end up being present in a desired chemical states. With that said, we declare that the third objective has been fulfilled.

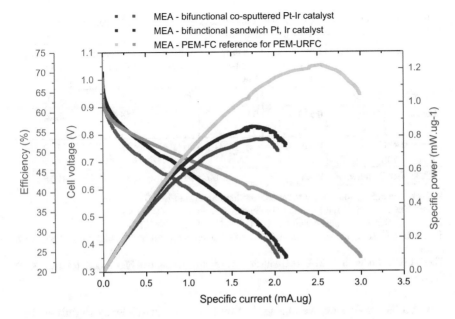

Fig. 3.70 Specific IV and power curves of **MEA—bifunctional sandwich-like Pt (bottom) Ir (top) catalyst** compared to **MEA—bifunctional co-sputtered Pt–Ir catalyst** and **MEA—PEM-WE reference for PEM-URFC**

3.5 Round-Trip Efficiency of PEM-URFC with Thin-Film Bifunctional Anode Catalyst

This final chapter compares and contrasts the round-trip efficiencies of our experimental thin-film low-loading bifunctional MEAs (within PEM-URFCs) and the referential dedicated single-purpose MEAs (within PEM-WEs and PEM-FCs). As it was stated in the introductory parts of the thesis, the main motivation for merging the PEM-WE and PEM-FC technologies together into the PEM-URFC is the potentially significant decrease of the price. Combining the aforementioned specialized systems into one universal device however inevitably leads to certain compromise regarding materials, PEMs or GDLs which is in turn reflected in lower round-trip efficiency. Obviously, the ideal trade-off is to maximize the reduction of noble metals without sacrificing much of the performance in individual regimes.

The table below compiles the noble metal loadings, WE and FC mode efficiencies, calculated using Eqs. (1.13) and (1.14), as well as round-trip efficiencies at 1 A cm^{-2} of combined single-purpose MEAs and bifunctional MEA which were tested and/or discussed throughout the thesis (Table 3.14).

The combination of referential single-purpose MEAs, with combined catalyst loading of 2.4 mg cm^{-2} yielded the round-trip efficiency of 40.02%. By using our optimized PEM-WE MEA, featuring 50 nm thin-film supported on TiC-based sub-

Table 3.14 Catalyst loadings, WE and FC efficiencies at 1 A cm^{-2} as well as round-trip efficiencies of MEAs at 1 A cm^{-2} discussed within the thesis

PEM-WE	PEM-FC	Loading PEM-WE [mg cm^{-2}]	Loading PEM-FC [mg cm^{-2}]	Efficiency PEM-WE [%]	Efficiency PEM-FC [%]	Roundtrip Efficiency [%]
WE ref.	FC ref.	1.6	0.8	87	46	40.02
Thesis WE	FC ref.	0.513	0.8	89	46	40.94
Thesis URFC co-sputtering		0.620		86	32	27.52
Thesis URFC sandwich		0.620		89	35	31.15

WE ref: The high-loading WE reference from literature (see Fig. 3.7)
FC ref: The measured high-loading FC reference (see Fig. 3.51, Table 3.11)
Thesis WE: The best experimental PEM-WE (see Fig. 3.49, Table 3.10)
Thesis URFC co-sputtering: PEM-URFC with the co-sputtered anode (see Figs. 3.55, 3.57, Table 3.12)
Thesis URFC sandwich: PEM-URFC with the sandwich-like anode (see Figs. 3.66, 3.69, Table 3.13)

layer on the anode side, we managed to push the overall efficiency slightly higher, while reducing the combined loading to 1.313 mg cm^{-2}. This alone is a very positive result, proving high application potential of thin-film low-loading catalysts on anode side of PEM-WE. Turning the attention to the bifunctional MEAs, we conclude that when merging the PEM-FC and PEM-WE anodes together in the right way (i.e. in stepwise sandwich-like manner), we are able to obtain very similar PEM-WE efficiencies as in case of single-purpose thin-film MEA. Admittedly, the efficiency in PEM-FC mode dropped compared to the high-loading reference. Yet considering that reducing the loading by more than one half (from 1.313 to 0.620 mg cm^{-2}) lowered the round-trip efficiency relatively moderately from 40.94 to 31.15%, we find such give-and-take very acceptable. One should not forget, that merging the systems in form of PEM-URFC leads to substantial cost savings not only through the catalyst reduction but also due to unification of hardware, which as was already shown in Fig. 1.10, represents at least at present time the majority of investments.

References

1. Hwang C, Ito H, Maeda T, Nakano A, Kato A, Yoshida T (2013) Flow field design for a polymer electrolyte unitized reversible fuel cell. ECS Trans 50:787–794. https://doi.org/10.1149/05002.0787ecst
2. Zhang D, Duan L, Guo L, Wang Z, Zhao J, Tuan WH, Niihara K (2011) TiN-coated titanium as the bipolar plate for PEMFC by multi-arc ion plating. Int J Hydrogen Energy 36:9155–9161. https://doi.org/10.1016/j.ijhydene.2011.04.123
3. Langemann M, Fritz DL, Muller M, Stolten D (2015) Validation and characterization of suitable materials for bipolar plates in PEM water electrolysis. Int J Hydrogen Energy 40:11385–11391. https://doi.org/10.1016/j.ijhydene.2015.04.155

4. Carmo M, Fritz DL, Mergel J, Stolten D (2013) A comprehensive review on PEM water electrolysis. Int J Hydrogen Energy 38:4901–4934. https://doi.org/10.1016/j.ijhydene.2013. 01.151
5. DOE Technical Targets for Hydrogen Production from Electrolysis. Energy Gov https://energy. gov/eere/fuelcells/doe-technical-targets-hydrogen-production-electrolysis (accessed 15 Nov 2017)
6. Ito H, Maeda T, Nakano A, Takenaka H (2011) Properties of Nafion membranes under PEM water electrolysis conditions. Int J Hydrogen Energy 36:10527–10540. https://doi.org/10.1016/ j.ijhydene.2011.05.127
7. Slavcheva E, Radev I, Bliznakov S, Topalov G, Andreev P, Budevski E (2007) Sputtered iridium oxide films as electrocatalysts for water splitting via PEM electrolysis. Electrochim Acta 52:3889–3894. https://doi.org/10.1016/j.electacta.2006.11.005
8. Sapountzi FM, Divane SC, Papaioannou EI, Souentie S, Vayenas CG (2011) The role of Nafion content in sputtered IrO_2 based anodes for low temperature PEM water electrolysis. J Electroanal Chem 662:116–122. https://doi.org/10.1016/j.jelechem.2011.04.005
9. Ma L, Sui S, Zhai Y (2008) Preparation and characterization of Ir/TiC catalyst for oxygen evolution. J Power Sources 177:470–477. https://doi.org/10.1016/j.jpowsour.2007.11.106
10. Sui S, Ma L, Zhai Y (2011) TiC supported Pt–Ir electrocatalyst prepared by a plasma process for the oxygen electrode in unitized regenerative fuel cells. J Power Sources 196:5416–5422. https://doi.org/10.1016/j.jpowsour.2011.02.058
11. Ma L, Sui S, Zhai Y (2009) Investigations on high performance proton exchange membrane water electrolyzer. Int J Hydrogen Energy 34:678–684. https://doi.org/10.1016/j.ijhydene. 2008.11.022
12. Huang J, Li Z, Zhang J (2017) Review of characterization and modeling of polymer electrolyte fuel cell catalyst layer: the blessing and curse of ionomer. Front Energy 11:334–364. https:// doi.org/10.1007/s11708-017-0490-6
13. Vielstich W, Lamm A, Gasteiger HA, Yokokawa H (2003) Handbook of fuel cells: fundamentals, technology, and applications. Wiley
14. Kúš P, Ostroverkh A, Ševčíková K, Khalakhan I, Fiala R, Skála T, Tsud N, Matolin V (2016) Magnetron sputtered Ir thin film on TiC-based support sublayer as low-loading anode catalyst for proton exchange membrane water electrolysis. Int J Hydrogen Energy 41:15124–15132. https://doi.org/10.1016/j.ijhydene.2016.06.248
15. van der Merwe J, Uren K, van Schoor G, Bessarabov D (2014) Characterisation tools development for PEM electrolysers. Int J Hydrogen Energy 39:14212–14221. https://doi.org/10.1016/ j.ijhydene.2014.02.096
16. Hoffman DW (1990) Intrinsic resputtering—theory and experiment. J Vac Sci Technol A Vac Surf Film 8. https://doi.org/10.1116/1.576483
17. Smith GC, Hopwood AB, Titchener KJ (2002) Electron inelastic mean free path for Ti, TiC, TiN and TiO_2 as determined by quantitative reflection electron energy-loss spectroscopy. Surf Interface Anal 33:230–237. https://doi.org/10.1002/sia.1205
18. Lu G, Bernasek SL, Schwartz J (2000) Oxidation of a polycrystalline titanium surface by oxygen and water. Surf Sci 458:80–90. https://doi.org/10.1016/S0039-6028(00)00420-9
19. Ottakam Thotiyl MM, Freunberger SA, Peng Z, Chen Y, Liu Z, Bruce PG (2013) A stable cathode for the aprotic Li–O_2 battery. Nat Mater 12:1050–1056. https://doi.org/10.1038/nmat3737
20. Albert A, Barnett AO, Thomassen MS, Schmidt TJ, Gubler L (2015) Radiation-grafted polymer electrolyte membranes for water electrolysis cells: evaluation of key membrane properties. ACS Appl Mater Interfaces 7:22203–22212. https://doi.org/10.1021/acsami.5b04618
21. Kúš P, Ostroverkh A, Khalakhan I, Fiala R, Kosto Y, Šmíd B, Lobko E, Yakovlev Y, Nováková J, Matolínova I, Matolín V (2019) Magnetron sputtered thin-film vertically segmented Pt–Ir catalyst supported on TiC for anode side of proton exchange membrane unitized regenerative fuel cells. Manuscript submitted to Int J Hydrogen Energy
22. Ostroverkh A, Johánek V, Dubau M, Kúš P, Khalakhan I, Šmíd B, Fiala R, Václavů M, Ostroverkh Y, Matolín V (2019) Optimization of ionomer-free ultra-low loading Pt catalyst for anode/cathode of PEMFC via magnetron sputtering. Int J Hydrogen Energy. https://doi.org/10. 1016/j.ijhydene.2018.12.206

23. Radev I, Topalov G, Lefterova E, Ganske G, Schnakenberg U, Tsotridis G, Slavcheva E (2012) Optimization of platinum/iridium ratio in thin sputtered films for PEMFC cathodes. Int J Hydrogen Energy 37:7730–7735. https://doi.org/10.1016/j.ijhydene.2012.02.015
24. Wang J, Holt-Hindle P, MacDonald D, Thomas DF, Chen A (2008) Synthesis and electrochemical study of Pt-based nanoporous materials. Electrochim Acta 53:6944–6952. https://doi.org/10.1016/j.electacta.2008.02.028
25. Fiala R (2017) Investigation of new catalysts for polymer membrane fuel cells. Dissertation Thesis, Charles University
26. Khalakhan I, Choukourov A, Vorokhta M, Kúš P, Matolínová I, Matolín V (2018) In situ electrochemical AFM monitoring of the potential-dependent deterioration of platinum catalyst during potentiodynamic cycling. Ultramicroscopy 187:64–70. https://doi.org/10.1016/j.ultramic.2018.01.015
27. Topalov AA, Katsounaros I, Auinger M, Cherevko S, Meier JC, Klemm SO, Mayrhofer KJJ (2012) Dissolution of platinum: limits for the deployment of electrochemical energy conversion? Angew Chemie—Int Ed 51:12613–12615. https://doi.org/10.1002/anie.201207256
28. Sugawara Y, Okayasu T, Yadav AP, Nishikata A, Tsuru T (2012) Dissolution mechanism of platinum in sulfuric acid solution. J Electrochem Soc 159:779–786. https://doi.org/10.1149/2.017212jes
29. El Sawy EN, Birss VI (2009) Nano-porous iridium and iridium oxide thin films formed by high efficiency electrodeposition. J Mater Chem 19:8244. https://doi.org/10.1039/b914662h
30. Cherevko S, Geiger S, Kasian O, Mingers A, Mayrhofer KJJ (2016) Oxygen evolution activity and stability of iridium in acidic media. Part 1—Metallic iridium. J Electroanal Chem 773:69–78. https://doi.org/10.1016/j.jelechem.2016.04.033
31. Cherevko S, Geiger S, Kasian O, Mingers A, Mayrhofer KJJ (2016) Oxygen evolution activity and stability of iridium in acidic media. Part 2—Electrochemically grown hydrous iridium oxide. J Electroanal Chem 774:102–110. https://doi.org/10.1016/j.jelechem.2016.05.015
32. Lee I, Whang C, Lee Y, Hwan G, Park B, Park J, Seo W, Cui F (2005) Formation of nano iridium oxide: material properties and neural cell culture. Thin Solid Films 475:332–336. https://doi.org/10.1016/j.tsf.2004.08.076

Chapter 4
Summary and Conclusions

The presented doctoral thesis falls within the very topical area of hydrogen economy. It revolves around proton exchange membrane water electrolyzers (PEM-WEs) and proton exchange membrane regenerative fuel cells (PEM-URFCs); more specifically it investigates the feasibility of using magnetron sputtering for deposition of thin-film, low-noble-metal-loading catalysts for the anode (oxidation) side of these electrochemical devices.

Following the first objective of the thesis, we set up the PEM-WE/PEM-URFC experimental cell and testing station. After experimenting with different arc-deposited protective coatings for anode end plate/current collector, we came to the conclusion that 5 μm thick layer of TiN is resistant enough to withstand highly corrosive conditions present on oxygen evolution reaction (OER) side of the cell. Thus, it was standardly used on anode side thereafter.

The second objective and core of the work focuses on preparation and characterization of thin-film anode catalyst for the PEM-WE. The motivation for this endeavor came from the current dire need of minimizing the noble metal content within the membrane electrode assembly (MEA) which is pivotal in respect to future broader commercialization of PEM-WE technology. The novelty of our approach lays in the use of magnetron sputtering for low-loading catalyst deposition.

It turned out that direct sputtering of 50 nm of metallic Ir on anode side of PEM does not lead to the desired OER activity nor stability. Scanning electron microscopy (SEM) analysis revealed that lateral expansion of PEM upon hydration ruptures Ir thin film which results in loss of its conductivity.

By sputtering Ir onto the Ti mesh instead, we were able to obtain stable performance; however, due to mesh's low surface the efficiency was far below what can be considered competitive with the state-of-the-art.

Notable improvement was achieved by coating a standard high-surface carbon paper gas diffusion layer (GDL) by 50 nm thin protective layer of Ti and consequently by 50 nm of Ir, forming gas diffusion electrode (GDE). Although the performance of such MEA was approaching values more typical for convention high-loading MEAs, this setup is not suitable for commercial applications since Ti protective coating of

© Springer Nature Switzerland AG 2019

P. Kúš, *Thin-Film Catalysts for Proton Exchange Membrane Water Electrolyzers and Unitized Regenerative Fuel Cells*, Springer Theses,
https://doi.org/10.1007/978-3-030-20859-2_4

carbon GDL cannot guarantee its long-term durability. Nonetheless, it was proven that Ti-coated carbon paper GDL can serve as a convenient high-surface substitution for more expensive Ti-based GDLs, at least during short-term R&D experiments.

It had been clear that in order to achieve high performance of thin-film catalyst, the use of high-surface support was inevitable. Keeping in mind that carbon materials are prone to corrosion, we turned our attention to the prospective corrosion-resistant alternatives. Nanoparticles of TiC seemed to meet the requirements for sufficient conductivity and electrochemical stability. As such, a major part of the thesis deals with novel Ir thin-film catalysts supported on TiC-based sublayer spread on anode side of PEM. Extensive comparative in-cell experiments and thorough characterization and comprehension of electrochemical phenomena within MEA resulted in determination of optimal parameters of support sublayer (i.e. its overall loading and Nafion content within) as well as in finding ideal thickness (i.e. loading) of Ir thin film in terms of absolute and specific (per unit mass of noble metal) performance. The ideal configuration of Ir/TiC anode catalyst was found out to be 0.1 mg cm^{-2} of TiC-based support material, containing 15 wt% of Nafion in respect to TiC, sputtered over by 50 nm of Ir.

Chemical analysis based on data from photoelectron spectroscopies (PES) with various information depths revealed that after electrochemical cycling in potentials typical for PEM-WE operation TiC does partially oxidize yet the effect is predominantly superficial and electrically conductive TiC state is still present. Hence it was proved that TiC-based sublayer is in principle capable of withstanding high anodic potentials and is therefore suitable support for OER catalyst.

Regarding the Ir catalyst itself, similarly performed electrochemical cycling showed that as-sputtered metallic Ir undergoes full potential-driven oxidation to IrO_2.

Finally, additional MEA optimizations were carried out, such as identifying the ideal PEM in terms of stability and conductivity as well as substituting Ti mesh anode GDL with sintered porous micro-grained Ti GDL. All of the above mentioned steps led to the design of a novel thin-film PEM-WE MEA capable of achieving very similar efficiencies as high-loading convention MEAs while relying on just a fraction of noble metal within.

The third and final objective of the thesis was to build on previous promising results and try to modify the novel thin-film anode PEM-WE catalyst in a way that would allow its operation on anode (OER/HOR) of PEM-URFC.

Our measurements showed that co-sputtering of Ir with Pt, which is necessary for catalyzing hydrogen oxidation reaction (HOR), onto the optimized TiC-based sublayer does not lead to ideal in-cell performances. Although performance drop in FC regime in comparison to the standard single-purpose reference was somehow expected due to the catalyst loading reduction, use of thicker PEM and inevitable replacement of several carbon-based components; the reason for inferior efficiency in WE mode compared to our previous MEAs with same parameters (except of Pt addition) was not clear.

PES analysis done after the co-sputtering of catalyst hinted formation of Pt–Ir alloy. Interestingly, after the electrochemical cycling the Ir did not fully oxidize to

IrO_2 but partially remained in a metallic state. Pt did not change its metallic state whatsoever. Unlike Pt though, which is active towards HOR in its as-deposited Pt^0 state, our experiments suggest that in case of sputtered iridium, full transition from Ir^0 to Ir^{4+} is needed for sufficient activity towards OER. The reason for this incomplete oxidation was explained using the electrochemical atomic force microscope (EC-AFM). It turned out that in contrast to the pure Ir layer which upon exposure to PEM-WE anodic potentials continuously oxidizes during which it also notably changes its topography, the Pt–Ir layer stays electrochemically and topographically more or less unaltered. Additionally, the cyclic voltammogram of Pt–Ir layer looks indistinguishable from pure Pt. Based on these results, we came to a conclusion that over the duration of co-deposition or shortly after introduction of high anodic potential the Pt–Ir alloy actually forms Ir-rich core/Pt-rich shell. As such we were able to see partially oxidized Ir under the skin of Pt, using PES with information depth of a couple of nanometers, yet the very surface sensitive cyclic voltammogram of Pt–Ir was shaped like pure Pt.

A clear difference between the behavior of pure Ir and Pt–Ir thin films was also evident from more bulk-sensitive energy-dispersive X-ray spectroscopy (EDX). While Pt–Ir spectra before and after electrochemical cycling matched each other and no significant oxygen peak emerged, pure Ir underwent truly thorough electrochemically induced oxidation which can be verified by a significant rise of oxygen signal within spectra.

Together the above mentioned findings indicated that due to the formation of core-shell structure and insufficient oxidation of Ir, the co-sputtering of Pt and Ir might not be the ideal way to prepare bifunctional thin-film anode catalyst for PEM-URFC.

Keeping this in mind we proposed an alternative preparation process which lead to sandwich-like Ir/TiC/Pt/PEM design of anode catalyst. We realized that in contrast to OER catalyst, the HER catalyst can indeed be so-to-say buried under the TiC-based sublayer without losing its activity due to the mass transport limitations. Convenient separation of catalysts for individual half-reactions thus prevented the electrochemical interplay in-between them and resulted in superior in-cell performance in both operational regimes. In fact, we obtained very similar WE efficiencies with sandwich-like PEM-URFC MEA as in case of single-purpose optimized PEM-WE MEA from the preceding part of the thesis.

Regarding the round-trip efficiency at 1 A cm^{-2}, the experimental sandwich-like thin-film PEM-URFC MEA yielded 31.15% in contrast to 40.02% achieved by combination of dedicated high-loading MEAs and in contrast to 40.94% given by the coupling of optimized thin-film PEM-WE MEA with high-loading referential PEM-FC MEA. The moderately lower round-trip efficiency of PEM-URFC MEA is however more than sufficiently redeemed by its notably lower noble metal content of just 0.620 mg cm^{-2} in comparison to 2.4 mg cm^{-2} of combined dedicated high-loading MEAs and 1.313 mg cm^{-2} of thin-film PEM-WE MEA coupled with high-loading referential PEM-FC MEA. On top of that, merging a water electrolyzer and fuel cell into one efficiently working device obviously leads to significant cost savings not only on catalyst but on overall hardware as well. Therefore, in light of presented promising results, we believe that further R&D in field of thin-film, low-loading PEM-URFCs is the step in right direction in context of cost optimization of electricity-H_2-electricity buffering cycle.

Summing up the findings of this dissertation, we have successfully proved that magnetron sputtering is a viable method for deposition of thin-film, low-loading catalysts for anode side of PEM-WE and PEM-URFC. Considering remarkably high utilization of deposited noble metal catalysts and stressing out wide industrial availability and cost-effectiveness of magnetron sputtering, we believe that applying herein described catalyst coated membrane (CCM) and MEA preparation methods might help in the endeavor to accelerate the commercialization of PEM-based hydrogen technologies.

Finally, we declare that all the objectives set in the introductory part of this thesis have been fulfilled.

Author's CV

Dr. Peter Kúš, Ph.D. (10. 10. 1989, Banská Bystrica, Slovakia)
Vincenta Hložníka 3542/8, 841 05 Bratislava, Slovakia
+420 721 072 210, peter.kus@mff.cuni.cz
ORCID ID: orcid.org/0000-0002-2246-4426

Education

2013–2018	Charles University, Faculty of Mathematics and Physics
	– *Ph.D. candidate (Physics of Surfaces and Interfaces)*
2011–2013	Charles University, Faculty of Mathematics and Physics
	– *Master's degree (Physics of Surfaces and Ionized Media)*
2008–2011	Charles University, Faculty of Mathematics and Physics
	– *Bachelor's degree (General Physics)*

Experience

2010–present	Member of a research group at the Department of Surface and Plasma Science (Faculty of Mathematics and Physics—Charles University)

© Springer Nature Switzerland AG 2019
P. Kúš, *Thin-Film Catalysts for Proton Exchange Membrane Water Electrolyzers and Unitized Regenerative Fuel Cells*, Springer Theses,
https://doi.org/10.1007/978-3-030-20859-2

- *Scientific work in the field of proton exchange membrane fuel cells and electrolyzers*
- *Focusing on development of nanostructured thin-film catalysts, utilizing complex multi target magnetron sputtering*
- *Experience in advanced morphology characterization by scanning electron microscopy and atomic force microscopy*
- *Experience in electrochemical cell testing*

2016 (5 months)   Fellowship at National Institute for Materials Science (Tsukuba, Japan)

- *Focusing on development of fuel cell cathode with low-loading of platinum*

2013 (1 month)   Cofely (GDF Suez)—Finance and Support Division intern
2012 (2 months)   Intern at the Department of Experimental Physics (Faculty of Mathematics, Physics and Informatics—Comenius University in Bratislava)

- *Focusing on preparation of thin films using DC magnetron sputtering*

2012 (1 month)   TopSoft BSB Ltd.—intern at the Department of Network Development

Research Grants

2018–present   TAČR #TG01010108
                Development of thin-film catalyst with low noble metal content for unitized regenerative fuel cells (Principal Investigator)
2017–present   GAČR #18-06989Y
                Novel materials for proton exchange membrane water electrolyzers (Co-Investigator)
2017–2018   GAUK #1016217
                Reversible hydrogen fuel cells based on thin-film nanostructured catalysts with minimum noble metal loading (Principal Investigator)
2016–2018   GAUK #897316
                Influence of fluorine doping on oxygen storage capacity of Rh/CeOxFy catalysts (Co-Investigator)
2014–2017   GAUK #236214
                Nanostructured gas sensors based on tin and cerium oxides, doped by noble metals (Principal Investigator)

Attended Workshops

2012 (1 week)   Kvant Ltd. workshop—Scanning electron microscopy, EDX, WDX (Tescan systems, Comenius University in Bratislava)

2011 (1 week)   Kvant Ltd. workshop—Atomic force microscopy (NT-MDT systems, Comenius University in Bratislava)

2006 (1 week)   Modern microscopy workshop, digital imaging and correct laboratory practice (Comenius University in Bratislava)

Language Skills

	Slovak (native language)
	Czech (fluent)
	English (fluent)
	German (intermediate)
2008	German Graduation (Level B)
2008	English Graduation (Level A)
2007	Certificate of Advanced English (ESOL)
2005	First Certificate in English (ESOL)
2004 (3 weeks)	Mountlands Language School (UK)
2003 (2 weeks)	English Language Course (ESE-Malta)

Additional Information

Computer literacy	MS Office, Origin
Specialized software literacy	KolXPD, Gwyddion, Inventor, EC-Lab
Experienced with methods/techniques	SEM, EDX, AFM, XPS, magnetron sputtering, electrochemical cell testing
Number of scientific publications	19 (April 2019)
Number of citations	88, h-index 5 (April 2019)
Talks at international conferences	6 (April 2019)
Posters at international conferences	8 (April 2019)
Supervised student projects	4 (April 2019)
Driving license (B)	

Interests

Traveling, Skiing, Sci-fi and Fantasy

List of Scientific Publications (April 2019)

Ostroverkh, A; Johánek, V; Dubau, M; Kúš, P; Khalakhan, I; Šmíd, B; Fiala, R; Václavů, M; Ostroverkh, Y; Matolín, V: Optimization of ionomer-free ultra-low loading Pt catalyst for anode/cathode of PEMFC via magnetron sputtering, Int. J. Hydrogen Energy, 2019. https://doi.org/10.1016/j.ijhydene.2018.12.206

Kot, M; Henkel, K; Naumann, F; Gargouri, H; Lupina, L; Wilker, V; Kus, P; Pozarowska, E; Garain, S; Rouissi, Z; Schmeißer, D: Comparison of plasma-enhanced atomic layer deposition AlN films prepared with different plasma sources, Journal of Vacuum Science & Technology A, 37 (2): Art. No. 020913 (11 pages), 2019. https://doi.org/10.1116/1.5079628

Kot, M; Kegelmann, L; Das, C; Kus, P; Tsud, N; Matolinova, I; Albrecht, S; Matolin, V; Schmeisser, D: Room-Temperature Atomic-Layer-Deposited Al2O3 Improves the Efficiency of Perovskite Solar Cells over Time, ChemSusChem, 11 (20): 3640–3648, 2018. https://doi.org/10.1002/cssc.201801434

Khalakhan, I; Choukourov, A; Vorokhta, M; Kúš, P; Matolínová, I; Matolín, V: In situ electrochemical AFM monitoring of the potential-dependent deterioration of platinum catalyst during potentiodynamic cycling, Ultramicroscopy, 187 (Apr): 64–70, 2018. https://doi.org/10.1016/j.ultramic.2018.01.015

Khalakhan, I; Waidhas, F; Brummel, O; Vorokhta, M; Kúš, P; Yakovlev, YV; Bertram, M; Dopita, M; Matolínová, I; Libuda, J; Matolín, V: Nanoscale Morphological and Structural Transformations of PtCu Alloy Electrocatalysts during Potentiodynamic Cycling, J. Phys. Chem. C, 122 (38): 21974–21982, 2018. https://doi.org/10.1021/acs.jpcc.8b06840

Gauter, S; Haase, F; Solař, P; Kylián, O; Kúš, P; Choukourov, A; Biederman, H; Kersten, H: Calorimetric investigations in a gas aggregation source, J. Appl. Phys., 124 (7): Art. No. 073301 (10 pages), 2018. https://doi.org/10.1063/1.5037413

Rednyk, A; Mori, T; Yamamoto, S; Suzuki, A; Yamamoto, Y; Tanji, T; Isaka, N; Kúš, P; Ito, S; Ye, F: Design of Active Sites on Nickel in the Anode of Intermediate-Temperature Solid Oxide Fuel Cells using Trace Amount of Platinum Oxides, ChemPlusChem, 83 (8): 756–768, 2018. https://doi.org/10.1002/cplu.201800170

Haviar, S; Chlupová, Š; Kúš, P; Gillet, M; Matolín, V; Matolínová, I: Micro-contacted self-assembled tungsten oxide nanorods for hydrogen gas sensing, Int. J. Hydrog. Energy, 42 (2): 1344–1352, 2017. https://doi.org/10.1016/j.ijhydene.2016.09.187

Khalakhan, I; Vorokhta, M; Kúš, P; Dopita, M; Václavů, M; Fiala, R; Tsud, N; Skála, T; Matolín, V: In situ probing of magnetron sputtered Pt-Ni alloy fuel cell catalysts during accelerated durability test using EC-AFM, Electrochim. Acta, 245 (10 Aug): 760–769, 2017. https://doi.org/10.1016/j.electacta.2017.05.202

Monai, M; Montini, T; Melchionna, M; Duchoň, T; Kúš, P; Chen, C; Tsud, N; Nasi, L; Prince, KC; Veltruská, K; Matolín, V; Khader, MM; Gorte, RJ; Fornasiero, P: The effect of sulfur dioxide on the activity of hierarchical Pd-based catalysts in methane combustion, Appl. Catal. B-Environ., 202 (Mar): 72–83, 2017. https://doi.org/10.1016/j.apcatb.2016.09.016

Khalakhan, I; Lavková, J; Matolínová, I; Vorokhta, M; Potin, V; Kúš, P; Václavů, M; Maraloiu, V-A; Kuncser, A-C; Matolín, V: Electrochemically shape-controlled transformation of magnetron sputtered platinum films into platinum nanostructures enclosed by high-index facets, Surf. Coat. Technol., 309 (15 Jan): 6–11, 2017. https://doi.org/10.1016/j.surfcoat.2016.11.017

Ostroverkh, A; Dubau, M; Johanek, V; Kus, P; Khalkhan, I; Vaclavu, M; Fiala, R; Ostroverkh, Y; Matolin, V: Optimization of Pt Catalyst for Anode/Cathode of PEMFC via Magnetron Sputtering, ECS Trans., 80 (8): 839–845, 2017. https://doi.org/10.1149/08008.0839ecst

Ostroverkh, A; Johanek, V; Dubau, M; Kus, P; Veltruska, K; Vaclavu, M; Fiala, R; Smid, B; Ostroverkh, Y; Matolin, V: Novel Fuel Cell MEA Based on Pt-C Deposited by Magnetron Sputtering, ECS Trans., 80 (8): 225–230, 2017. https://doi.org/10.1149/08008.0225ecst

Kúš, P; Ostroverkh, A; Ševčíková, K; Khalakhan, I; Fiala, R; Skála, T; Tsud, N; Matolin, V: Magnetron sputtered Ir thin film on TiC-based support sublayer as low-loading anode catalyst for proton exchange membrane water electrolysis, Int. J. Hydrog. Energy, 41 (34): 15124–15132, 2016. https://doi.org/10.1016/j.ijhydene.2016.06.248

Ostroverkh, A; Johánek, V; Kúš, P; Šedivá, R; Matolín, V: Efficient Ceria–Platinum Inverse Catalyst for Partial Oxidation of Methanol, Langmuir, 32 (25): 6297–6309, 2016. https://doi.org/10.1021/acs.langmuir.6b01316

Khalakhan, I; Fiala, R; Lavková, J; Kúš, P; Ostroverkh, A; Václavů, M; Vorokhta, M; Matolínová, I; Matolín, V: Candle Soot as Efficient Support for Proton Exchange Membrane Fuel Cell Catalyst, Fuel Cells, 16 (5): 652–655, 2016. https://doi.org/10.1002/fuce.201600016

Kettner, M; Ševčíková, K; Duchoň, T; Kúš, P; Rafaj, Z; Nehasil, V: Morphology and CO Oxidation Reactions on Anion Doped CeOXFY/Rh(111) and CeOX/Rh(111) Inverse Catalysts, J. Phys. Chem. C, 120 (47): 26782–26792, 2016. https://doi.org/10.1021/acs.jpcc.6b07431

Monai, M; Montini, T; Melchionna, M; Duchoň, T; Kúš, P; Tsud, N; Prince, KC; Matolin, V; Gorte, RJ; Fornasiero, P: Phosphorus poisoning during wet oxidation of methane over Pd@CeO2/graphite model catalysts, Appl. Catal. B-Environ., 197 (15 Nov): 271–279, 2016. https://doi.org/10.1016/j.apcatb.2015.10.001

Vorokhta, M; Khalakhan, I; Vaclavu, M; Kovacs, G; Kozlov, SM; Kus, P; Skala, T; Tsud, N; Lavkova, J; Potin, V; Matolinova, I; Neyman, KM; Matolin, V: Surface composition of magnetron sputtered Pt-Co thin film catalyst for proton exchange membrane fuel cells, Appl. Surf. Sci., 365 (Mar): 245–251, 2016. https://doi.org/10.1016/j.apsusc.2016.01.004

Printed in the United States
By Bookmasters